Stonaker, Frances Benson
Famous mathematicians. [1st ed.] Philadelphia, Lippincott [1966]
118p. 22cm.

1. Mathematicians. I. Title.

FAMOUS MATHEMATICIANS

FAMOUS MATHEMATICIANS

Frances Benson Stonaker

J·B·Lippincott Company
PHILADELPHIA *and* NEW YORK

To J. L. S.

PRINTED IN THE UNITED STATES OF AMERICA
Library of Congress Catalog Card Number 66-15159
FIRST EDITION

ACKNOWLEDGMENTS

I am particularly indebted to Dr. Paul F. Baum, of the Department of Mathematics, Princeton University, for his patient and generous assistance in clarifying the mathematical concepts which appear in this book. Others who were kind enough to offer information, anecdotal material, or comments include Herbert S. Bailey, Jr., Clarence D. Chang, Mildred G. Goldberger, Judith A. Milgram, Malcolm Skolnick, Charles M. Terry, and Frederick H. van Doorninck, Jr.

Special thanks are due Joseph L. Bolster III, Joseph A. Lopez, and Mary Jenkins Snedeker, my young readers, for their thoughtful and constructive suggestions.

Finally, without the constant encouragement, support, and understanding of my husband, Joseph L. Stonaker, this book could never have been completed.

CONTENTS

	INTRODUCTION	*ix*
I	EUCLID	*3*
II	ARCHIMEDES	*12*
III	ARYABHATTA; AL-KHWARISMI	*24*
IV	DESCARTES	*35*
V	NEWTON	*45*
VI	LAGRANGE	*57*
VII	GAUSS	*67*
VIII	GALOIS	*79*
IX	VON NEUMANN	*91*
X	WIENER	*103*
	INDEX	*116*

INTRODUCTION

MANY PEOPLE think of mathematics as a tiresome and never-ending series of sums to be added or amounts to be divided, and imagine a mathematician to be a kind of human computer. Nothing could be further from the truth! A close look at the "queen of the sciences" reveals that the mathematical world abounds in beautiful and intriguing problems, many of which are of universal significance. It is not surprising, then, that the inhabitants of this world are fascinating people to know.

Much of the work of mathematicians comes under the heading of "pure" mathematics, which can be defined as the investigation of mathematical theories and ideas. The development of the theory of equations by Galois is an example of pure mathematics at its best. By contrast, "applied" mathematics is concerned with the application of mathematical theories to problems in other branches of science. Wiener's work in cybernetics offers an outstanding example of this phase of mathematics.

All great mathematicians, whether ancient or modern, pure or applied, have brought to their science an intense dedication and zest for its challenges. It is the author's hope that, through the biographies of these men, the reader will sense the excitement of the mathematical adventure, and may, perhaps, be tempted to partake of its delights.

FAMOUS MATHEMATICIANS

I

EUCLID

PROUD SHIPS rocked contentedly in the Alexandrian harbor, their holds bulging with Oriental treasure. Now and then a breath of warm Egyptian air brushed the bronze torches at the wharf's edge, sending shafts of light darting among inky waves. In the midnight sky, constellations followed one another in leisurely pursuit as they moved toward the Mediterranean horizon.

This night, the drowsy silence of the city was broken by festive sounds. The columned palace of Ptolemy, King of Egypt, echoed with laughter and gaity. In its great hall, tunic-clad courtiers talked animatedly, while servants rushed to fill goblets.

"My esteemed friend!" At the sound of Ptolemy's regal voice, the courtiers' gossiping quieted; eyes turned toward the carved throne. The king was addressing his guest of honor, Euclid (YOO-klid).

"I have begun to read your new work on geometry, good Euclid," Ptolemy continued. "An excellent endeavor! Still, I find that this admirable book is somewhat long and difficult. What shorter path to mathematical knowledge exists for Ptolemy, who must devote his life to his kingdom?"

The short, bearded man on his right gazed steadily at the king. Slowly he shook his head. "Sire," he said firmly, "even for you, there is no royal road to geometry."

Gasps ran through the room. How dared Euclid address the master of Egypt in this fashion?

The king looked thoughtfully at Euclid. "I respect your opinion, learned mathematician," he finally answered. "It is our good fortune that you have come to Alexandria."

Ptolemy spoke truly that night. It was the good fortune of Alexandria, and of the world, that Euclid wrote and thought about mathematics. Every student and scholar of geometry from ancient times until now has been indebted to him.

Not much is known of Euclid's early life. It is believed that he was born in the ancient coastal city of

Tyre, in what is now Lebanon, in about 330 B.C. A few years before, the great Greek general Alexander had captured Tyre. Perhaps Euclid's parents came to the city then; at any rate, he was of Greek descent.

As a young man, Euclid crossed the Mediterranean to Athens, Greece, where he studied at the famous Academy. This center of learning, which had been founded by the philosopher Plato, influenced much of Euclid's later work. Plato had recommended the study of astronomy, arithmetic, geometry, and music harmony to his followers, and these are the four subjects on which Euclid concentrated in his writings.

While Euclid was studying in Athens, important changes were taking place in Greek history. Alexander the Great, who had conquered most of the then-known world for Greece, had died. His empire was divided among three of his generals. One of these generals, Ptolemy, who had been governor of Egypt, now became its king. Ptolemy chose Alexandria as the new capital of Egypt.

Alexandria was named after Alexander himself. In his travels, the mighty conqueror had come across its site on the Mediterranean coast, just west of the mouth of the Nile River. Realizing that it offered a good trading location and an excellent harbor, Alexander had begun construction of a city there.

Ptolemy decided to make Alexandria a renowned

center of learning as well as a wealthy world trading port. He founded the famous "Musaeon," or University of Scholars, and invited Greek thinkers in every field of study to come and teach there. One of those invited was Euclid.

It is interesting to imagine Euclid's feelings when he arrived in Alexandria and saw the University. The first of such institutions, its description resembles that of a modern university. We are told that it had lecture rooms, laboratories, museums, library facilities, and living quarters. Its famous library was for centuries the largest in the world, and was said to have had 600,000 papyrus rolls. The leading historians, mathematicians, astronomers, and poets of the time found in Alexandria an ideal center of Greek culture.

Euclid began his masterpiece, the *Thirteen Books of Elements,* in Alexandria. Although much of its reasoning is probably Euclid's, the *Elements* is a collection, organization, and explanation of all mathematical writing known at that time. An examination of this work tells us a great deal, not only about Euclid's talents as teacher and mathematician, but also about Greek mathematics from 500—300 B.C.

An unusual feature of the *Elements* is the form in which the mathematical material was presented. Euclid began his work with a list of "general understandings"—statements which he considered to be obviously true, and which were the foundation of all other ideas proposed in the *Elements*. The first five statements, called "axioms," are as follows:

1. Things which are equal to the same thing are also equal to one another.
2. If equals be added to equals, the wholes are equal.
3. If equals be subtracted from equals, the remainders are equal.
4. Things which coincide with one another are equal to one another.
5. The whole is greater than the part.

All of Euclid's mathematical propositions proceed in gradual sequence from these five axioms, plus five additional statements which he wrote about lines, circles, and angles. Each idea unfolds as a consequence of a previous one, in a step-by-step, logical order. This unique arrangement became a model for the method of reasoning used by mathematicians from that time until now.

As the title suggests, the *Elements* contains thirteen sections, called "books." Much of the material deals with geometry, that branch of mathematics which is the study of space. For example, *Book I* considers triangles, angles, and squares. One of its propositions is the famous Pythagorean theorem, named for another Greek mathematician, Pythagoras. It states that the square of the longest side (hypotenuse) of any right triangle is equal to the sum of the squares of the other two sides. In other words, if the length of the longest side is multiplied by itself, the result will equal the sum of the lengths of each other side multiplied by itself. (A right triangle can be recognized by the perfect L-shaped angle opposite the hypotenuse.)

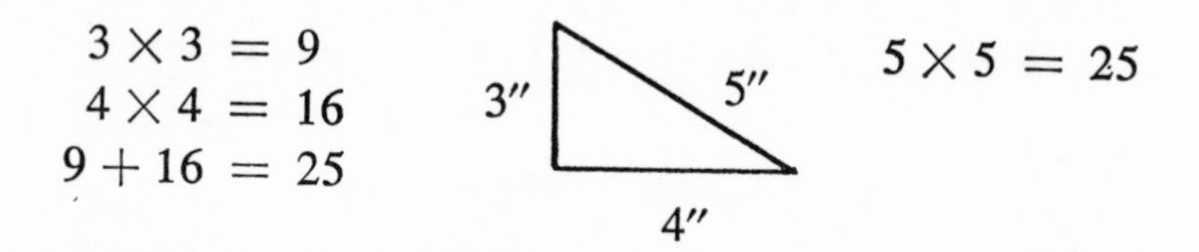

Book II contains several interesting problems about straight lines and segments (parts) of these lines, as well as further study of triangles. Circles and the measurement of angles highlight *Book III* of the *Elements*. Much of the material in this book is included in American high school geometry books; it has traveled far in time and place from Alexandria!

Euclid included instructions for constructing geometric figures in several books of the *Elements*. One section explains the construction of regular polygons of 3, 4, 5, 6, and 15 sides, using only straightedge (ruler) and compass. Another describes the method of constructing such solid figures as the cube or pyramid.

Solid geometry, which is concerned with three-dimensional figures, was not neglected by Euclid. *Book XI* lays the foundation for its study. The theory of lines and planes in space is also developed in this section.

Euclid did not concentrate on geometry alone in his work. Two books of the *Elements* contain information on the theory of proportions (the relationship between the size and amount of things), while three others concern themselves with the theory of numbers. For example, *Book IX* contains the following statements:

A sum of even numbers is even.
A sum of an even number of odd numbers is even.
A sum of an odd number of odd numbers is odd.
Even less even is even.
Even less odd is odd.
Odd less odd is even.
Odd less even is odd.
Odd times even is even.
Odd times odd is odd.

The *Elements* is most of all a tribute to Euclid's outstanding teaching ability. He arranged the material so that the easiest work was presented first, and his explanations and proofs are clear and orderly. No sooner was the *Elements* completed than the scholars of Alexandria hailed it as a great work of mathematics.

The path traveled by the *Elements* in its journey from 300 B.C. to the twentieth century was a complicated one. For hundreds of years it circulated through the Greek world, widely used as the standard mathematics textbook. In the eighth century A.D. it was translated into Arabic, under the scholarly eye of Caliph Al-Mamun, a medieval ruler of Bagdad. This Arabic manuscript was translated into Latin in 1120 by an Englishman, Adelard of Bath, and other Latin translations followed.

A printed edition of the *Elements,* made at Venice in 1482, increased its European circulation. Finally, in 1570, Euclid's great work made its appearance in English. More than one thousand editions of the *Elements* have been printed since that time; it has been the basis of geometry teaching for centuries.

The fame of the *Elements* has partly obscured Euclid's other mathematical works. One of these, the *Data,* is an important early book dealing with algebra. (Algebra is the branch of mathematics which uses special symbols, such as letters, to express relationships between numbers.) Some of the propositions in the *Data* explain how to determine unknown amounts; others are concerned with substitution.

Another work, *On the Divisions of Figures,* concerns itself with the problem of dividing geometric figures, such as rectangles or triangles. Unfortunately,

Euclid's remaining mathematical works have been lost.

Euclid spent many fruitful years as mathematics professor in Alexandria. He was known as an honest, humble man. In addition to his works of mathematics, he was the author of books on perspective, reflection by mirrors, theory of music, and elementary astronomy.

Euclid always insisted that knowledge was worth acquiring for its own sake. A student who was just beginning to study geometry once asked him, "What shall I gain by learning these things?"

Euclid looked at him sadly. "Here is threepence," he replied, "since you must make gain of what you learn."

It is through the *Elements* that Euclid's name lives in history. Students throughout the centuries all over the world have learned geometry from him. We are fortunate to be his pupils!

⋙ II ⋘

ARCHIMEDES

"SIRE, ROMAN SHIPS approach! They will soon be upon us! The gods alone can save us from the might of the Roman sword!"

King Hieron gazed quietly at the panting young soldier before him, then turned toward the sun-flecked sea. A cluster of toylike ships, dancing on the bright Mediterranean, inched closer to the fortressed city of Syracuse. The king's lips curved in a slow smile.

"I shall give no orders now," he stated calmly. "We shall wait."

Silence followed his words. The tiny vessels grew larger, sharper. Soon the clatter of shields and spears rang over the water; the stabbing noon sun haloed hundreds of helmets. Mercilessly the fleet drew closer.

Suddenly King Hieron made a sharp gesture. At once the air was filled with a hail of huge boulders. Strange, awkward machines catapulted stones and lead blocks at the enemy. Robot-like claws of iron reached over the fortress walls, seized the curved ship prows, and dashed them into the sea's depth.

The Roman advance tottered, then fell back in fright and confusion. As King Hieron watched the battered fleet depart, his smile broadened.

"There is our deliverer!" he cried, gesturing toward an aged man standing obscurely beside one of the catapults. "Noble Archimedes! Once again you have proved yourself the glory of Syracuse!"

Archimedes (ahr-ki-MEE-deez) was the greatest mathematician, engineer, and physicist of ancient times. He was born in the mighty Greek colony of Syracuse, on the east coast of Sicily, in 287 B.C. His father, the astronomer Phaedius, was noted for his research on the diameters of the sun and moon. Except for a period of study in Alexandria, Archimedes spent most of his life in the city of his birth.

Syracuse was a bustling outpost of Greece in the third century B.C. The beauty of Greek civilization was mirrored in her pillared temples and public buildings. Her mighty shipping fleet roamed the Mediterranean at will, while at home the fine Syracusan harbor invited the trade of Mediterranean neighbors. No wonder the Roman legions coveted this renowned port!

Archimedes showed little interest in Syracusan affairs, preferring to spend his time writing and making mathematical discoveries. However, he was often interrupted in his work by frantic pleas for advice from King Hieron, ruler of Syracuse. As often as the king would present him with a problem which seemed impossible to solve, Archimedes would counter with an astounding answer or an amazing invention.

For example, King Hieron once had a magnificent ship, the *Syracosia,* built for Ptolemy of Alexandria. Skilled craftsmen had equipped the vessel with every luxury known at that time. No sooner was it finished, however, than the builders realized that it was too large to launch! The king, wringing his hands in despair at the thought of such an expensive failure, sent for his brilliant subject.

Archimedes listened good-naturedly to the king's tale of woe. Returning to his study, he designed an apparatus of cogwheels worked by an endless screw. This launching mechanism, a kind of giant lever, could be operated by one man alone.

As he worked, Archimedes realized the possibility of moving even larger objects in this manner. "Give me a spot where I can stand, and I shall move the earth!" he exclaimed.

King Hieron was beside himself with joy at Archimedes' success. As he singlehandedly launched the ship, the Syracusan monarch proclaimed, "From this day forward, Archimedes will be believed, no matter what he says!"

Another of Archimedes' interesting inventions was a water-removing device called the Archimedean screw, or Cochlias. It consisted of a tube which was open at both ends and bent into the form of a spiral, like a corkscrew. The device was first used in Spain to pump water from silver mines. Later, sailors of ancient times took advantage of Archimedes' idea, and bailed ships with their version of his screw. The invention was amazingly successful in Egypt, too, where it was used to irrigate parched crops during the dry season, and to drain fields after the Nile's annual floods. Proof of the durability of the Archimedean screw is the fact that it is still being used by Egyptian farmers, twenty-two centuries later.

Archimedes himself considered these inventions to be unimportant, since his only real interest was mathematics itself. Plutarch, the Roman biographer, remarked of Archimedes, "Although these discoveries had brought him the fame of superhuman sagacity, he did not want to leave behind any writing on these subjects; he considered the construction of instruments, and, in general, every skill which is exercised for its practical uses, as lowbrow and ignoble, and he only gave his efforts to matters which, in their beauty and their excellence, remain entirely outside the realm of necessity."

Like many later mathematicians, Archimedes often left his meals untouched when he was involved in an absorbing problem. His usual chalkboards were sanded floors or smooth earth, but if these were not available, Archimedes would draw his diagrams on the cold ashes of an old fire, or even on his own oil-rubbed skin!

The mathematical writings of Archimedes are masterpieces which prove his genius. Beginning with an explanation of the plan to be followed, they proceed in order through his propositions, eliminating every unnecessary detail. "It is not possible to find in geometry more difficult and troublesome questions or proofs set out in simpler and clearer propositions," Plutarch comments.

One of Archimedes' works, *Measurement of the Circle,* is concerned with plane geometry. Here Archimedes set forth the classical method of computing π. This strange symbol is the Greek letter pi (pronounced "pie"), which in ancient times was thought to have an approximate value of 3. For example, the mathematical sentence $C = \pi d$ means that the circumference (C) of a circle is equal to about three times the diameter (d) of the circle. (The circumference is the distance around the circle—its edge; the diameter is the distance from one side of the circle through the center to the other side.)

Archimedes was anxious to find a more accurate figure for π. Carefully, brilliantly, he developed proofs to show that the value of π lay someplace between $3\frac{1}{7}$ and $3\frac{10}{71}$. (Eight centuries later, on the other side of the earth, the Chinese mathematician Tsu Ch'ung-chih discovered that $\frac{355}{113}$ was an even more accurate estimate of π's value.)

A lucky discovery made it possible for twentieth century mathematicians to understand more fully the reasoning behind Archimedes' proofs. In 1906, J. J. Heiberg, a Danish expert on ancient scholarly writings, was invited to Constantinople to inspect a papyrus manuscript from a monastery in Jerusalem. A study of the manuscript convinced Dr. Heiberg that it really consisted of two separate works, one on top of the other. The original text, in Greek, was a work of mathematics which included many diagrams. However, this writing had been rubbed off centuries later by monks, and replaced with a religious text.

Luckily, Dr. Heiberg was able to restore and decipher most of the original text. At first the results seemed disappointing; though the author of the work was evidently Archimedes, the material was already familiar. Then came a long section which was new, at least to modern mathematicians. A brief study revealed the exciting truth: Dr. Heiberg had found the priceless *Method,* the long-lost work of Archimedes in which he explains how he arrived at many of his theorems.

In the *Method,* Archimedes shows himself to be a true mathematician in his insistence on mathematical reasoning as the basis for all scientific thinking. "Certain things first became clear to me by a mechanical method, although they had to be demonstrated by geometry afterwards because their investigation by (mechanical) method did not furnish an actual demonstration," he writes. "But," he continues, "it is of course easier, when we have previously acquired . . . some knowledge of the questions, to supply the proof than it is to find it without any previous knowledge."

The *Method* sets forth Archimedes' great discovery that the volume of the circumscribed cylinder is 1½ times as great as the volume of the sphere. (To get an idea of this statement, think of a cylinder, such as a tin can, as high as it is wide. Now picture a hollow ball just as high and just as wide. The tin can would hold 1½ times as much liquid as the ball.)

—the tin can, or cylinder
—the ball, or sphere

Such a discovery was outstanding in Archimedes' time, and his thinking on this problem was faultless. In a later work on solid geometry, *On the Sphere and the Cylinder,* Archimedes gave mathematical proofs of its correctness.

Archimedes must have enjoyed thinking about huge quantities. *The Sand Counter,* a work on arithmetic, deals with the estimation and notation of large numbers, such as the number of grains of sand in an imaginary, sand-filled universe! It is said that the mathematician once posed a problem dealing with white, black, yellow, and dappled cows and bulls. "The Cattle Problem," as it has come to be called, requires as the least possible answer a number of more than 206,500 digits!

Archimedes' other writings include *On Spirals, On Floating Bodies, On Levers,* and *On Mirrors,* the last two of which are lost. He sent letters to his friend, Conon, an Alexandrian astronomer, about many of his theories. The proofs were usually omitted from these letters, because, as he remarked, he didn't want to keep from other mathematicians the pleasure of finding out the answers for themselves. (Sometimes, as a joke on one or two of them, he slipped in a false proposition, "so that those who pretend to have discovered everything themselves, without supplying proofs, may fall into a trap by asserting to have found something which is impossible.")

The scientific interests of Archimedes seem to have included astronomy, though we know little about his accomplishments in this field. However, one endeavor for which he was admired was the construction of a revolving planetarium. This open sphere was run internally; a single switch set in motion models of the sun, the moon, and the five planets known at that time. Each rotated and revolved in exact precision. "When the sphere was set in motion, one could, at every turn, see the moon rise above the earth's horizon after the sun, just as it occurs every day; and then one saw how the sun disappeared and how the moon entered into the shadow-cone of the earth with the sun on the opposite side," wrote Cicero, the great Roman author, who had seen the sphere in operation.

In his belief that the earth is at the center of the universe, Archimedes agreed with the commonly accepted but erroneous view of the astronomer, Ptolemy. One writer tells us that the mathematician calculated the distance from the earth to the moon, from the moon to Venus, to Mercury, to the sun, to Mars, to Jupiter, to Saturn, and even to the nearest stars. Unfortunately, Archimedes' work, *On the Making of the Spheres,* which probably included such information, is lost.

Once Archimedes played the part of a detective, with humorous results. King Hieron had given a certain amount of gold to a goldsmith with instructions to fashion it into a crown. When the crown was delivered, Hieron suspected that the man had kept some of the gold and substituted silver in its place. Since the weight of the crown equaled that of the gold, however, he could not prove his suspicion.

As usual, the king turned to Archimedes for help. Several days later, as Archimedes stepped into his bath, he noticed that the water level rose, and that his body seemed lighter, as he sank into the water. Forgetting where he was, he leaped out of the bath and ran to the palace, crying, "Eureka! Eureka!" (which means, "I have found it!")

What Archimedes had found, and what he later proved by experiments, was that when equal weights of gold and silver are weighed in water, they no longer appear equal, because the silver, having a greater volume, displaces more water. So ended "The Case of the Dishonest Goldsmith"!

Archimedes was an old man when the Romans began their siege of Syracuse. For three years his strange mechanical devices repelled them. Particularly effective were the "scorpions," small catapults which kept the Roman army under a stinging shower of rock. The aging mathematician was even supposed to have designed a giant mirror, in the form of a hexagon surrounded by polygons, which focused the sun's rays on the invading ships and set them on fire.

Marcellus, the Roman general, was infuriated by these tactics. "Shall we not make an end of fighting against this geometrical giant, who uses our ships to ladle water from the sea, who has ignominiously driven off our battering rams, and who by the multitude of missiles he hurls at us all at once outdoes the hundred-armed giants of mythology?" he thundered at his engineers.

The Roman aides looked helplessly at one another. "If our men only see a rope or a piece of wood extending beyond the walls, they take flight, exclaiming that Archimedes has invented a new machine for their destruction," one explained.

In desperation, Marcellus ordered a blockade, against which even Archimedes was powerless. Food grew scarce, and eventually the starving city surrendered.

Upon entering Syracuse, Marcellus ordered that the life of Archimedes be spared. Hours later, as the mathematician sat at home drawing a geometric diagram in the sand, a Roman soldier entered his room. "Don't step on my work," cautioned Archimedes.

Angered, the soldier drew his sword and killed him.

In recognition of the mathematician's genius, the Roman leader had a splendid monument built to honor Archimedes. Engraved on the stone, as he had wished, was a carved representation of a sphere and a cylinder, a reminder of his greatest mathematical achievement. A far more lasting and magnificent memorial to Archimedes exists in his own works, however. They have been a challenge to mathematicians for centuries.

III

ARYABHATTA; AL-KHWARISMI

If a modern mathematician were given a page of arithmetic problems in poem form and instructed to write all his answers in verse, he would probably protest that mathematics and poetry make an impossible combination. Yet, centuries ago, Hindu mathematicians of India recorded all their ideas and discoveries in poems. In doing so, they made an important contribution to our number system.

One of these poetic Hindu mathematicians was Aryabhatta (ahr-yuh-BUH-tah). He was born near Patna, India, on the Ganges River, in about A.D. 476. At that time Patna was a thriving center of Indian culture. Deep moats surrounded its turreted walls, which in turn enclosed well-planned roads, colorful bazaars, theaters, and inns. A magnificent palace, plated with gold and silver, rose at the center of the city. Its imperial park boasted ornamental lakes and unusual birds from many lands.

The Gupta emperors ruled India at this stage of her history, and their elephant army protected Indian subjects from the threat of Hun invasions. Freedom and civilized rule were Gupta trademarks; theirs was an age of learning and art.

As Aryabhatta grew up, he became familiar with the fanciful tales of India. One of these, his favorite, may have stimulated his interest in numbers. According to legend, Prince Gautama (Buddha) once asked Prince Dandapani for the hand of his daughter in marriage. So many young men had made this request that Dandapani set up a series of contests in writing, wrestling, archery, running, and swimming. Buddha was victorious in each.

Last came an arithmetic contest. A great mathematician began to question Buddha:

"How do the numbers beyond the *koti* continue by hundreds?" (The *koti*, according to one arithmetic book, is one hundred times one hundred thousand.)

Buddha was prepared for this question. "One hundred *kotis* are called *ayuta,* one hundred *ayutas* are *niyuta,* one hundred *niyutas* are *kankara,* one hundred *kankaras* are *vivara* . . ."

On and on Buddha recited, through twenty-three stages. Finally he paused briefly. "I have come to the end of the series," he stated. "However, eight similar series follow."

Not surprisingly, Buddha's knowledge of these extraordinary numbers brought him the desired marriage.

Aryabhatta was one of the few Hindus of his day whose interest in mathematics was encouraged. He belonged to the ruling class of Hindus, called Brahmans. The members of this group had for years considered themselves superior to other Hindus. By Aryabhatta's time only the Brahmans were allowed to receive a mathematical education. This practice caused Hindu mathematical progress to decline for centuries.

Aryabhatta's contribution to mathematics, however, was significant. His book, the *Aryabhattiya,* is one of the earliest preserved Indian mathematics texts. It is divided into four sections, all in verse. The first and second deal with a system of number writing and mathematics; the last two discuss the reckoning of time and the heavenly bodies.

The dedication of the section on mathematics is

a colorful one. It reads: "Having paid homage to Brahma, the Earth, the Moon, Mercury, Venus, the Sun, Mars, Jupiter, Saturn, and the constellations, Aryabhatta, in the City of Flowers, sets forth the science venerable." The chapter goes on to consider familiar topics, including the areas of triangles and circles, the relationship of the circumference of the circle to its diameter, and fractions.

The concern of Aryabhatta with a system of number writing was to have far-reaching results. Hindu mathematicians used the ancient Brahmi numerals, which are the ancestors of our 1, 2, 3, etc. But until the time of Aryabhatta, the idea of place value, or position, was unknown in India. The same number of units, tens, and hundreds was represented by different numerals. Thus 555 would be written in Hindu numerals. (In our 555, each 5 gets its value from its place, or position, in relation to the other numerals.)

Realizing that numerals were awkward and unpoetic, the Hindus substituted related words in order to keep their verses rhythmical. For example, they wrote "moon" instead of one, because there is only one moon. "Wings" or "eyes" took the place of two. (In the same way, we might substitute "planets" for nine, or "fingers" for ten.)

Aryabhatta took the first step towards a positional

number system with his strange "syllable-numbers." The vowels in each syllable of these words indicated units, tens, or hundreds. For example, Aryabhatta wrote the number 3336 as:

ca	ya	gi	yi
6	3	3	3

beginning with the units on the left, as all Hindus did.

The word is like a code. Syllables which contain "a" are units and tens; syllables containing "i" are hundreds and thousands. Each consonant has a different value: c = 6, y = 30, g = 3. We might translate cayagiyi as 6 ones, 30 ones, 3 hundreds, 30 hundreds.

Aryabhatta's method may seem an awkward, confusing one, yet it paved the way for the work of his pupil, Bhaskara. This mathematician, who lived about A.D. 520, introduced an improved system which *was* positional. In Bhaskara's poetry, syllables which stood for 3 could also stand for 30 or 300, depending on their place in the word. Bhaskara's system has a zero as well, probably borrowed from Greek mathematics.

The *Aryabhattiya* includes rules for finding the square and cube root of numbers. One verse instructs: "In the method of inversion, multipliers become divisors and divisors become multipliers; addition become subtraction and subtraction, addition." (As an example, if $6 + 4 = 10$, then $10 - 4 = 6$.)

In the role of astronomer, Aryabhatta taught that what seems to be the daily rotation of the heavens is really the rotation of the earth around its axis. This idea was considered a daring one in that century, and for many centuries afterwards. He also calculated the time at which eclipses occur, and the length of the earth's shadow.

It is fortunate that Aryabhatta and other Hindu mathematicians of his time realized the importance of writing numbers in a simple way. The development of a positional number system with only ten digits (0-9) was a giant step in the history of mathematics. But how did this great Hindu contribution reach Europe? That is another story . . .

One evening, three centuries after the time of Aryabhatta, a caravan moved wearily across heat-faded sands towards Bagdad. Its uneven pace and the glazed eyes of its travelers betrayed a futile search for an oasis. On the western horizon, the desert seemed about to burst into flames, then was lost in shadows as the sun slipped from its grasp.

Suddenly the silent dusk was shattered by the wild cries and slashing scimitars of desert bandits! As the caravan members scrambled for protection, the marauders struck. Spice sacks tumbled, knives ripped open cartons of precious silks, and rough hands seized delicate gems. Then, as quickly as it had begun, the fierce attack ceased.

Minutes passed. Slowly the travelers stirred, then started uncertainly towards the ruined goods, muttering sadly at their loss.

Only one caravan member remained motionless. As he watched the others move through the wreckage, a relieved smile came over his face. The package hidden under his robes remained intact. Desert pirates had no use for so priceless a treasure!

The caravan reached Bagdad several days later. As it entered the city, the sights and sounds of the marketplace rose in greeting. Shouting vendors and bawling camels thronged labyrinthine alleys, where perfumers, money changers, silk weavers, and booksellers traded. The odor of cinnamon and saffron rose from dim shop cubicles; veiled women gossiped in the shadows. Above mud brick dwellings and the palm-fringed Tigris River, the relentless sun blazed.

As the caravan lurched to a halt, the same passenger slipped quickly from his camel. Weaving his way through twisting, narrow streets, he came at last to an inconspicuous doorway. At his knock, the door opened, and he was ushered inside to a glad greeting.

"Peace be with you, al-Khwarismi! We are glad that you have returned safely. What news do you bring?"

Al-Khwarismi (ahl-khwahr-IZ-mee) bowed deeply, then strode across the ornately carpeted floor. Twisting flames flickered in the incense-filled air, illuminating magnificent tapestries which hung on each wall. The dusty traveler bowed again before speaking.

"On you be peace, and the mercy and blessings of God, oh noble Caliph," he answered. "Allah has granted me a successful journey. I have brought you a treasure beyond price to adorn our House of Wisdom."

As al-Khwarismi drew a package from the folds of his cloak, the Caliph leaned forward intently. The wrappings fell aside, revealing a small tablet covered with cryptic symbols. The Caliph hestitated, then looked up quizzically.

"A treasure, my friend? Of what value are these strange markings?"

Al-Khwarismi smiled. "Great Caliph, before you lies the number system of the Hindus. Its economical use of a handful of numerals is a treasure of mathematical simplicity. With your aid, this achievement may soon be known throughout the Arab world."

Fortunately for al-Khwarismi, the mighty Caliph

Al-Mamun welcomed new ideas in every field of learning. His "House of Wisdom" included a scientific academy, an observatory, and a public library. Generous royal support was given to the scholarship of doctors, lawyers, musicians, poets, and scientists.

Under Al-Mamun, members of all religions enjoyed complete freedom of worship and belief, and religious and legal discussions were encouraged. The Caliph's messengers traveled the Mediterranean coast and beyond, seeking scholarly information from other peoples; a core of translators remained in Bagdad to work on the gathered material.

Al-Khwarismi was one of the Caliph's finest mathematical scholars, and, indeed, one of the great men of medieval science. His real name was Muhammed ibn Musa; the name by which we know him comes from his birthplace Khwarism, or Khiva, east of the Caspian Sea, where he was born in 750.

The influence of al-Khwarismi on the mathematics of this period was enormous, chiefly because he not only recognized the value of the Hindu numerals, but also wrote a small book about them.

What a traveler that book became! In its Arabic form, it circulated through the Moslem world, which at that time included Spain. Several centuries later, the work was translated into Latin, probably by the same English monk, Adelard, who translated the *Elements*. Few Europeans used the odd symbols at first; they preferred the familiar, awkward Roman numerals.

Eventually, however, an Italian mathematician, Fibonacci, published an arithmetic book in which he stated his preference for the Hindu-Arabic numerals, as they had begun to be called. The book was a great success, and encouraged the spread of the new system. In the twelfth century, a German arithmetic book included the numerals in a multiplication table whose form would be easily recognized by twentieth century students.

An interesting passage from al-Khwarismi's work explains the use of the zero. He writes: "When in subtraction nothing is left over, then write the little circle, so that the place does not remain empty. The little circle has to occupy the position, because otherwise there would be fewer places, so that the second position might be mistaken for the first."

The "little circle" was called *"sifr"* (meaning "empty") by the Arabs. It was translated into Latin as *zephyrum*, then in Italian was shortened to *zero*.

Another important work of al-Khwarismi was his *Calculation of Integration and Equation*. This text was translated into Latin and used as the chief mathematics book in European universities until the sixteenth century. It introduced the word *al-jabr,* or algebra, into general use.

In his spare time, al-Khwarismi compiled astronomical and trigonometrical tables, and joined other Arabic scholars in writing a geographical encyclopedia for Al-Mamun.

Luckily for the future of mathematics, Aryabhatta and al-Khwarismi joined hands across the centuries in a most effective way. Their thinking is echoed in every classroom around the world. Through their development and use of the decimal number system, these two mathematicians made a lasting contribution to the history of mathematics.

IV

DESCARTES

THE MUFFLED ROAR of cannons shook the flimsy barracks walls with dull persistence. Inside, a lone soldier lay on his wooden cot, staring thoughtfully into space. After a few minutes, he was distracted by a fly, marching soundlessly across the cracked ceiling. The soldier watched the insect as it headed for a corner.

Suddenly he sat up with a start. If the distance from each point on the fly's path to each of the nearest two walls were known, then the fly's path could be described mathematically. This insect might provide the key to the problem which had been plaguing him for months!

René Descartes' (day-KAHRT) fly did indeed prove to be the unwitting solution to an intriguing puzzle. The young man had seen in the insect's path a possible clue to a new method of approaching geometry. Descartes seized this slender hint and fashioned it into a brilliant discovery—one which placed him among the giants of modern mathematics.

Born near Tours, France, in 1596, Descartes grew up in that colorful region of fish-filled rivers and fairytale castles. As a boy, his health was poor, and he was seldom able to go fishing and exploring with his friends. Instead, he spent much of his time reading and questioning his parents about topics which interested him.

Descartes' father, a wealthy estate owner and member of the local parliament, realized that his son's exceptional curiosity was a sign of unusual intelligence. He decided to send the boy to La Flèche, a nearby boarding school, to be educated.

Because young René was so frail, the school's headmaster allowed him to rest each morning until he felt able to study. Descartes continued this practice for years. "The only way to do good work in mathematics and to preserve one's health is never to arise until inclined to do so," he once remarked.

Descartes seems to have been a successful student at La Flèche. Much of the school's curriculum was devoted to the study of Latin and Greek, and the boy learned these ancient languages with ease. Nevertheless, his rigorous schedule made him long for occasional relaxation, and in 1612, after taking his last examinations, Descartes set out for Paris with several friends.

For a while, the young men led a merry life in Paris, spending money recklessly. Gradually, however, Descartes became bored by such an aimless existence. About this time he met two former schoolmates from La Flèche who had moved to the city to embark on mathematical research. Their keen interest in mathematics proved to be contagious, and, for the next two years, Descartes led a life of uninterrupted mathematical study.

By 1616, most of Europe seemed like an armed camp, for religious wars had broken out across the continent. Descartes began to hear exciting stories about the military campaigns led by William of Orange, a Dutch prince. Stirred by the thought of adventure, he set out to join the army of Holland.

A few months later, as he strolled through a Dutch town, Descartes spied a poster on the wall of an ancient inn. The new recruit's curiosity was aroused. Hailing a passerby, he asked him to translate the Dutch words on the poster.

"Gladly," agreed the man, who happened to be a

Dutch scholar, "if you will attempt the geometry problem it poses!"

Descartes nodded with interest. No sooner had the scholar finished translating the poster than Descartes set to work. Within a few hours he had arrived at the correct solution of the problem, which had baffled mathematicians from all over Europe. So enjoyable was this experience to the young soldier that he decided to devote his life to mathematical pursuits.

Mathematics had developed slowly in medieval Europe. At the beginning of the seventeenth century, multiplication was considered quite difficult, and division was attempted only by skilled mathematicians. The plus (+) and minus (—) signs so familiar to us were just beginning to be widely used, and the equals sign (=) had only been introduced in 1540, by an English mathematician named Recorde.

While Descartes was taking part in a campaign near the Danube River in 1619, a curious occurrence took place which influenced his development as a mathematician. He tells us that he had a series of strange dreams: in the first, he saw himself blown by evil winds from school and church towards something of great strength. The second dream pictured him watching a terrifying storm with calm, scientific interest, realizing that its fury could not hurt him. In the third dream, he saw himself reciting a poem—"What Way of Life Should I Follow?"

Descartes interpreted these dreams as an explanation of the secrets of nature and a call to explore nature through mathematics. He immediately began work on a volume which was to become his life's work, and which was to have a profound influence on later scientific thought.

By 1621, Descartes had realized that mathematics offered him a far more challenging adventure than military campaigns. Resigning his commission, he spent the next five years traveling through Germany, Denmark, Holland, Switzerland, and Italy. Descartes was an unusual tourist; instead of sightseeing, he concentrated on mathematical study and discussion.

On one trip, in northern Europe, Descartes nearly lost his life. His ship's cutthroat crew had observed his suitcases bulging with important-looking papers, and concluded that the mathematician was a wealthy gentleman. They planned to knock him unconscious, steal his wallet, and heave him into the sea.

Luckily, Descartes overheard their plot. Striding courageously to the deck, sword in hand, he confronted the startled sailors. "Take me at once to the nearest port," he demanded, "or I will sharpen my blade on your necks!" The crew hurriedly obeyed, and the mathematician was soon out of danger.

Soon after that incident, Descartes had another chance to prove himself a swordsman. A rude fellow had insulted a lady who was with Descartes, and the mathematician challenged him to a duel on the spot. With a few deft strokes of his sword, Descartes forced the man to his knees in apology. "I will not take your life," roared Descartes, "chiefly because I wish to spare this lady the sight of your death!"

Descartes returned to Holland in 1628 to study philosophy and mathematics in earnest. There he roamed from town to town, experimenting and writing. He carried on a busy correspondence with great scientific thinkers from all over Europe, channeling letters through a former teacher so that he might be alone to think. Many subjects interested him at this time—chemistry, medicine, anatomy, meteorology, astronomy, optics—and he studied each of them. For a while he even attempted to construct optical instruments, and was quite successful.

It was during this second period in Holland that Descartes finished his scientific masterpiece, *Discourse on the Method of Rightly Conducting the Reason and Seeking Truth in the Sciences.* Although it was published in 1637, the *Discourse on the Method* had occupied eighteen years of Descartes' life. The time had been well spent.

"How do we know anything?" Descartes had often asked himself. "If we can't say definitely that we know anything, how will we ever find out anything?"

After years of puzzling and thinking, Descartes concluded that the scientific method offered the only answer. He felt that if a scientist experimented carefully and then applied mathematical reasoning to the results, his findings could be considered accurate. The *Discourse on the Method* set forth his ideas on this subject.

The most important section of this work, from a mathematical point of view, is an appendix entitled *Geometry*. The appendix contains the basis of modern analytic geometry—the method of solving geometry problems which Descartes had begun to dream of years before, and which was a milestone in mathematics.

In *Geometry,* Descartes unfolded a method which used algebraic formulas to describe geometric figures. This startling new approach beautifully complemented the geometry of Euclid. Descartes' discovery of analytic geometry was hailed as one of the most brilliant feats in mathematical history.

Descartes' other contributions to mathematical thought include important work in the theory of equations. (An equation is a mathematical sentence which expresses equality between two amounts.) He also simplified the writing of algebraic terms, and set forth a classification of curves.

Descartes' work on the physical universe, *The World,* set forth the unusual theory that the sun is the center of a huge whirlpool of matter, and that the planets which float in this whirlpool are swept around like tiny sailboats in a river current. Though this theory seems strange to our way of thinking, it did attempt to explain the patterns of the universe by the mechanical laws which apply on earth.

In 1646, Descartes received an invitation to tutor Queen Christine of Sweden. The nineteen-year-old queen had read of Descartes' mathematical theories, and decided that he should be her instructor in mathematics. She dispatched a special ship to bring the great man to Sweden.

Christine was no ordinary young lady. In addition to being a talented and independent ruler, she was a skilled horseback rider and huntress. She ate and slept little, and had a habit of reading for hours in her unheated library in the middle of winter.

Descartes responded to the queen's invitation with interest. He had heard that Christine was an eager pupil, and he looked forward to teaching her. Chanute, the French ambassador to Sweden, had offered Descartes a room in the embassy, and the mathematician planned to tour the forested countryside during his Swedish visit.

Poor Descartes! He was not prepared for the ferocious winters of Sweden, nor for Christine's strange whims. The strong-willed young queen demanded that Descartes arrive at the palace every morning at five o'clock to start the day's lessons. While snow blew through the open library windows, Christine happily discussed mathematics with her shivering tutor.

For a time Descartes tried to make up the lost hours of sleep by resting during the day. Soon, however, Christine decided that he would make a perfect chairman of the Royal Swedish Academy of Sciences. She thoughtlessly scheduled meetings of the Academy every afternoon, much to the despair of the exhausted mathematician.

After several years of this unhappy combination of cold, overwork, and lack of sleep, Descartes' lungs became seriously inflamed. He died in Stockholm in 1650, at the age of fifty-four. His body lies in the Pantheon in Paris, the final resting place of France's great men.

"Science may be compared to a tree," Descartes once commented. "Metaphysics is the root, physics is the trunk, and the three chief branches are mechanics, medicine, and morals." It might be added that only through the careful nourishing of great thinkers such as Descartes can the tree of science thrive.

V

NEWTON

"Watch Isaac's face when he comes through that door! He won't know what happened to him!"

The speaker, a tall boy about twelve years old, smirked unpleasantly as he nudged the student beside him. A small, curious band had formed around them in the school hall, and as the classroom door opened, all eyes turned towards the slight, rather shabby boy who emerged.

"Isaac!" The young pupil, turning, caught the full blow of of the larger boy's heavy boot. As he clutched at his stomach with a painful gasp, his attacker whirled and fled, leaving behind an echo of mocking laughter.

Isaac straightened uncomfortably, then stared down the deserted passageway, his face pale with fury. "That bully has tricked me once too often," he thought angrily. "*I'll* plan the next surprise."

In the following weeks, Isaac Newton (NOO-t'n) spent most of his spare time in his father's barn. Then, one afternoon, he crossed the school yard and confronted his enemy. "I'm challenging you to a fair fight right now," he stated firmly.

The sneer which broke across the larger boy's face changed abruptly to a grimace of pain at Isaac's first blow. The other students watched in amazement as the bully was driven to his knees with several well-timed punches. "You win, you win," he finally shouted, cringing before Isaac's stinging fists.

A wave of cheers which rose from the crowd was checked as the school's headmaster strode into view. Walking to the spot where Isaac had the older boy pinned to the ground, he addressed him sternly.

"Well, Isaac! Surely you can put your time to better use! Since you seem to have become the school's leader in fighting, perhaps you might now try to become its top scholar!"

Several of the onlookers snickered behind their hands. Isaac Newton a good student? That was impossible! He was always inventing toys or writing in his notebook when he should have been studying.

Isaac did not laugh, however. He hesitated a moment, then stood and faced the headmaster respectfully. "Sir, I accept your challenge," he answered. Then, picking up his scattered books, he started purposefully back to his classroom.

Young Isaac's decision was a momentous one. He not only rose to the top scholastic position in his school, but he eventually became the greatest mathematician in England's history, and one of the most brilliant scientists who ever lived.

Isaac Newton was born on Christmas Day, 1642, in a village in Lincolnshire, England. His mother later remarked that he was so tiny at birth that he could have fitted into a quart jar. Isaac's father, who was a farmer, hoped that his son would help to manage the family fields when he grew older.

Few who knew Isaac as a boy guessed that he had an unusual mind. His school grades had never been above average, and his teachers often complained of his inattention and daydreaming. The villagers looked on his habits of reading for hours at a time and of keeping a record of things which interested him as merely eccentric.

An early clue that hinted at Newton's genius, however, was his talent as a boy inventor. Some of his inventions were quite practical—sundials, a wooden clock that kept accurate time, several working waterwheels. Others must have been planned for fun. Isaac once figured out a way to build a kite with an enclosed lantern. When it was completed, he flew it at night in the village fields, laughing mischievously at the villagers' tales of ghostly demons hovering in the sky.

Isaac's most extraordinary invention was a small mill which mechanically ground wheat into flour. Harnessing a mouse to keep the mill going, he watched with delight as the tiny animal nibbled at the results of its labor.

When Newton was fifteen, his father died. This family tragedy placed the young man in the role of farm manager, one in which he had little interest. Fortunately, however, Isaac's uncle had noticed that his nephew often slipped away from his farm chores to read, and realized that the boy's talents lay in other directions. At his urging, Newton's mother allowed her son to enroll at Cambridge, one of England's great universities.

At Cambridge, for the first time in his life, Newton was exposed to the thinking of mathematicians before him. His introduction to this new world was accidental; while browsing through a town bookshop one afternoon, he came across a copy of Euclid's *Elements*. Newton's scanty background in mathematics could hardly have prepared him to understand the

Elements, but he was curious to learn about Euclid's ideas. Amazingly, Newton found the work easy reading, and was fascinated by it. He returned to the store, bought Descartes' *Geometry,* and quickly mastered that difficult book.

After that, science and mathematics occupied most of Newton's time. Though he supported himself during this period by working as a servant, the young man soon became a brilliant mathematical scholar. At the end of his first two years at Cambridge, his professor, Dr. Isaac Barrow, resigned his position in Newton's favor, thus publicly recognizing his student's astounding ability.

In 1664, when Newton was twenty-two years old, the Great Plague struck England. The appalling sanitary conditions of that century offered it little resistance, and the disease swept through the country, killing thousands swiftly and silently. Fearing infection, the university closed its doors, and professors and students alike returned to their homes.

Newton decided to spend this extended vacation developing theories which had interested him for some time, and, as a result, the next two years were among the most productive of his life. From 1664-1666, he invented the calculus, discovered the law of universal gravitation, and prepared his findings on the spectrum. This fantastic output of work laid the foundation for all his later theories, and so exhausted him that he became physically ill.

Newton's invention of the calculus was the most astounding mathematical achievement of the seventeenth century. It proved to be a brilliant method of analyzing problems which had long defied solution. As the mathematician John von Neumann has observed, "The calculus was the first achievement of modern mathematics, and it is difficult to overestimate its importance."

The creation and development of the calculus was a work of supreme genius. Nevertheless, Newton was quick to recognize his debt to the earlier work of Descartes, Kepler, and Galileo. "If I have seen a little farther than others, it is because I have stood on the shoulders of giants," the mathematician once modestly remarked.

Newton spent the next twenty years of his life lecturing at Cambridge and continued his mathematical work. So advanced was his thinking that mathematicians struggled for fifty years to understand what he produced during this period. Finally, in 1684, he began writing his masterpiece, the *Mathematical Principles of Natural Philosophy*.

The *Principia,* as it is called, is actually a summation of Newton's discoveries about the physical world. In his introduction to the first section, he stated that he intended to apply mathematics to the phenomena of nature, especially motion. He went on to formulate his three laws of motion, the third of which ("To every action there is an equal and opposite reaction") is essential to the understanding of rocket power and jet propulsion.

The main sections of the *Principia* deal with the motion of bodies in free space, motion in a resisting medium, and the application of the first section's findings to the solar system. Newton's treatment of these matters is magnificent, and the *Principia* has probably had more influence on scientific thinking than any other published work.

Once Newton had begun the *Principia,* he thought of little else until it was finished. His concentration seems unbelievable; he wrote for eighteen or nineteen hours daily, ignoring meals, sleeping little. He started writing the moment he awoke each day, and often sat for hours on the edge of his bed, working in his nightclothes.

In addition to the *Principia,* Newton wrote *The Universal Arithmetic,* which substantially advanced the theory of equations. His other works include writings on the calculus, plus brilliant investigations of curves, optics, and analytic geometry.

The works of Newton were published long after being written, and probably would never have been put into print except for the urging of the mathematician's friends. Newton himself dreaded the publication of his books, since he was constantly being called upon to explain their contents. "I see a man must either resolve to put out nothing new, or become a slave to defend it," he once remarked. And later he confided unhappily, "I was so persecuted with discussions arising out of my theory of light, that I blamed my own imprudence for parting with my quiet to run after a shadow."

Newton's reluctance to publish his writings had one very unfortunate result. Although he had invented the calculus in 1666, he had not bothered to publish a description of it until 1693. However, a German mathematician named Leibniz had, in the meantime, come to the same conclusions as Newton. His findings, published in 1684, were immediately hailed in Germany as a mathematical discovery.

When Leibniz's work reached England, a storm of controversy broke; accusations were hurled by friends of both mathematicians. Fanned by hot tempers, the quarrel became serious, and communication between English and German mathematicians was eventually broken. Ironically, the only winner in this dispute was France; her mathematicians, using material from both Newton and Leibniz, perfected the calculus and advanced mathematics in their own country.

Many biographers speak of Newton's fierce hatred of tyranny and love of justice. The mathematician once represented Cambridge in a dispute with King James II, who had demanded that the university award a degree to an unqualified member of his court. Newton refused to grant the king's order, and urged his colleagues to refrain from signing the compromise which was offered. "An honest courage in these matters will secure all, having law on our sides," he argued.

Newton seems to have been the original absent-minded professor. At his home, visitors restrained their laughter at the sight of two holes which he had cut in his front door, one for his cat and another for her kitten. It is said that the mathematician once dismounted from his horse to lead it up a steep hill, and that, as he climbed, his thoughts naturally turned to mathematics. It was not until he reached the top that he realized that the animal had wandered off, leaving Newton with an empty bridle!

One evening, Newton's guests discovered even more convincing proof of his forgetfulness. He had left the dinner table to get a bottle of wine from his cellar. Halfway there, the mathematician realized that he had forgotten his errand. He glanced around, puzzled, and noticed his chapel robe lying on a chair. "That's it! I'm late for service!" he thought with relief, and rushed off in the direction of the locked church!

Newton never married. He was said to have been engaged to a girl from his village when he was nineteen years old, but she later married someone else. He lived comfortably, however, for shrewd business investments made him quite wealthy. His work left him little time for hobbies, though he did have an interest in theology and chemistry.

In 1699, the English government decided to honor Newton by appointing him Master of the Mint. Though this belated recognition was hardly worthy of Newton's genius, the mathematician took his new job quite seriously. His duties were to reform and supervise the coinage of England, and he carried them out superbly. Nor was this his only contribution to the field of public service. Twice he acted as the representative of Cambridge University in Parliament, and in 1714, he advanced ocean navigation through his report to the House of Commons on determining longitude at sea.

Newton always enjoyed solving challenge problems in mathematics. Not surprisingly, he was second to none in the speed and brilliance of his answers. One evening, for example, he arrived home to find a friend waiting with two problems which had baffled the mathematicians of Europe for nearly a year.

Newton tackled the problems with delight, and, within several hours, found the solution to both of them. Then, mischievously, he mailed the answers to the mathematician Bernoulli without revealing his identity. As soon as Bernoulli read the anonymous solutions, however, he exclaimed, "Ah! I recognize the lion by his paw!"

After a long, full life, Newton died in London at the age of eighty-five, and was buried at Westminster Abbey. His genius had advanced the science of mathematics beyond the wildest dreams of the sevententh century. Leibniz, his rival, paid him this deserved tribute, "Taking mathematics from the beginning of the world to the time when Newton lived, what he did was much the better half." And the witty poet Alexander Pope wrote,

> "Nature and Nature's laws lay hid in night,
> God said, 'Let Newton be,' and all was light."

Newton himself, however, remained the humble scientist, aware of his own limitations. "I do not know what I may appear to the world," he wrote, "but to myself, I seem to have been only like a boy playing on the seashore, now and then finding a smoother pebble or a prettier shell than ordinary, while the great ocean of truth lay all undiscovered before me."

VI

LAGRANGE

AN ELEGANT CARRIAGE rattled down the country lane and came slowly to a halt beside the little roadside market. As the owner's wife caught sight of the vehicle's occupant, she smiled greedily. Such a fine gentleman! Surely he would heap the coachman's arms with fruit—and fill her pockets with sous—before the hour was out. She bustled to meet him.

The man who alighted from the coach glanced over the produce in a leisurely way, then turned to the peasant woman. "I wish to spend a few sous for an assorted basket of fruit," he remarked pleasantly. "Can you gather such a mixture?"

Reddening slightly, the woman began to stammer.

"Oh, sir, of course I would be honored . . . but you see . . . there is no way to measure such a mixed basket."

The gentleman stared at her incredulously. "No way to measure . . . ? But surely you have a scale, or some standard that is used. How do the people of the neighboring farms make such a selection?"

The peasant woman looked even more embarrassed. "Our closest neighbors live in the next district," she explained earnestly, "and their way of measuring is different from ours. But sir," she continued somewhat fearfully, "accept any fruit you wish to take. It is not my intention to anger such a noble gentleman as you."

Her customer seemed to have forgotten she was there. "Different measures!" he repeated in an unbelieving voice. "No standard! How can the people of France tolerate such an inefficient system?" Without another word, he climbed back into his carriage, shouted "Paris" to his astonished coachman, and disappeared down the lane, leaving the stunned woman in his wake.

If the departing customer had been an ordinary one, that might have been the end to the situation. However, the gentleman was none other than Joseph Louis Lagrange (la-GRANZH), one of the greatest mathematicians of the eighteenth century. And not the least of his contributions to mathematics was a remarkable solution to the peasant woman's dilemma.

Lagrange was born in the year 1736 in Turin, Italy. His father, a cavalry officer, was French; his mother, Italian. Eleven children were born to the Lagranges, but Joseph was the only one who lived.

Although Lagrange's father was at one time a wealthy man, he spent most of his money in unsuccessful business ventures. Lagrange considered this a stroke of luck. "If I had inherited a fortune, I should probably not have cast my lot with mathematics," he once commented contentedly.

As a schoolboy, Joseph was fascinated by the colorful legends and history of Greece and Rome. Although his classical studies included the works of Euclid and Archimedes, the young pupil showed little interest in mathematics at first. One day, however, he chanced to read an essay by Newton's friend Halley on the superiority of the calculus over early Greek geometry. The work impressed Lagrange so deeply that he decided to enroll at Turin University and concentrate on mathematics.

Within a short time, the young student's mathematical ability was the talk of the university. His amazed teachers, realizing that Lagrange was far too advanced for their classes, suggested that he join the faculty. As a result, he became a professor of mathematics at Turin at the age of sixteen!

Lagrange combined his new teaching duties with work in advanced mathematical research. As time went on, he decided to seek a mature opinion of his progress, and sent the results of his labors to the Swiss mathematician, Euler. The older man recognized Lagrange's genius at once.

Luckily for Lagrange, Euler was a generous man who believed in encouraging talented young mathematicians. He immediately speeded publication of Lagrange's work on the calculus of variations, and waited to publish his own theories on the same topic so that the younger man would receive the full honor of discovery. Later, in his writings, Euler explained and praised Lagrange's contribution to the mathematics of the time.

The grateful Lagrange more than justified Euler's faith in him. Between the years 1764 and 1788, he won five prizes offered by the French Academy of Sciences. One was awarded for his brilliant solution to the question of why the same side of the moon always faces the earth. Other problems which he attacked and solved concerned the motion of the moon, comets, and the planet Jupiter.

When Lagrange was thirty years old, the King of

Sardinia rewarded him for his achievement in mathematics with a trip to Paris and London. The journey nearly killed him, for he became seriously sick after attending a banquet in Paris. Although Lagrange had enjoyed the opportunity of discussing his theories with the leading mathematicians there, the illness dampened his enthusiasm for Paris.

No sooner had Lagrange returned to Turin than he received a letter from another king, Frederick the Great of Prussia. This proud monarch had heard of Lagrange's brilliance through Euler, and he wrote, "The greatest king in Europe wishes to have at his court the greatest mathematician in Europe." Flattered by such praise from the mighty Frederick, Lagrange accepted the offer.

Frederick immediately appointed Lagrange as director of the mathematical section of the Berlin Academy, a position Lagrange occupied for twenty years. His German colleagues, who were at first suspicious of the intruder, came to appreciate his considerate behavior and intellectual modesty. "I do not know," was Lagrange's reply to any mathematical problem about which he was in doubt. Then curiosity would take over, and he would tackle the troublesome question with keen interest.

Lagrange's method of working on a problem was somewhat different from that of other mathematicians. Before he began a paper, he thought it out, step by step, in great detail. When at last he wrote down his ideas, there was seldom need to make a single correction. He once confided the secret of his method to another mathematician: "As I am not pressed and work more for pleasure than from duty, I am like the great lords who build: I make, unmake, and remake, until I am passably satisfied with my results, which happens only rarely."

Precise workmanship and daring originality mark Lagrange's mathematical publications. His most splendid contributions were in the areas of equations, number theory, and the calculus of variations. The *Analytical Mechanics,* which Lagrange began when he was nineteen years old, has been described as a "scientific poem." It included general equations of motion which are known as "Lagrange's equations."

In 1786, Frederick the Great died. The interest in science and mathematics which the Prussian king had encouraged dwindled, and Lagrange began looking for a more hospitable atmosphere. Once again a king came to his aid—this time, Louis XVI of France. The young monarch invited Lagrange to become a member of the French Academy.

Upon his arrival in Paris, Lagrange met the king and his famous queen, Marie Antoinette, and was given an apartment in the Louvre, which later became France's greatest museum. Almost immediately, however, he was overcome by a strange fit of depression, during which he lost all enthusiasm for mathematics. At scientific meetings, he would gaze vacantly out the window, speaking rarely, listening halfheartedly. For the second time, it seems, Paris had a bad effect on Lagrange. Fortunately this period was a short one, and although the mathematician refused even to open his newly published *Analytical Mechanics,* he did find consolation in studying new subjects: languages, the history of religion, medicine, and botany.

An event of great historical importance finally roused Lagrange from his unhappy state. In July, 1789, the people of Paris stormed the Bastille and released the prisoners who were being held there by the king's orders. This blow against tyranny signaled the start of the French Revolution.

Although Lagrange was in sympathy with the French people's desire for freedom, he was aghast at their bloody revenge on the nobility, and took no part in it. He did, however, agree to serve the new government as president of a commission for the reform of weights and measures. For centuries, each French district had had its own measurement system, and the resulting chaos moved the government to seek Lagrange's assistance. The choice was a wise one; under his leadership the commission established the metric system, which is used today by all scientists, and is the official system of measures in hundreds of countries throughout the world.

The members of Lagrange's committee decided to draw up new units of length, weight, and volume to be used by all Frenchmen. They began by calculating the distance from the North Pole to the equator, and then took as their unit of length an amount equal to one ten-millionth of that distance. The new unit of length was called the *meter*, which is equal to 39.37 inches, or approximately 3 yards and 3 inches. The *gram* was chosen as the basic unit of weight, and the *liter* as the basic unit of volume.

A stroke of genius prompted Lagrange to insist on 10 as the metric system's base. This common-sense choice, which was approved by the commission, makes conversion from one amount to another a simple matter of multiplying or dividing by 10. Prefixes are used to indicate different amounts, as the chart below shows:

$$1 \text{ meter} = 39.37 \text{ inches}$$

$$\frac{1}{10} \text{ meter} = \text{decimeter (dm)}$$

$$\frac{1}{100} \text{ meter} = \text{centimeter (cm)}$$

$$\frac{1}{1000} \text{ meter} = \text{millimeter (mm)}$$

10 meters = decameter (dcm)
100 meters = hectometer (hm)
1000 meters = kilometer (km)

The same framework applies to weight and volume.

Use of the metric system spread rapidly through France and Europe. England never adopted it, unfortunately, nor did the United States. Both these countries still cling to the awkward British-American system of measures, although in 1866 the United States Congress passed a law allowing Americans to use the metric system, and recently several Congressmen have submitted bills urging its adoption by the United States.

Lagrange made another significant contribution to French mathematical progress in 1797, when he became the first professor of the newly opened *École Polytechnique* (Technical School) in Paris. The mathematician had not taught for many years, and his students' education had been interrupted by the Revolution. Nevertheless, Lagrange immediately drew up an excellent mathematics program, and before long his superb teaching attracted the best scientific minds of France. The École Polytechnique survived and prospered; today it is one of France's great schools.

Probably one of the reasons for Lagrange's return

to good spirits at this stage of his life was his marriage to the young daughter of an astronomer with whom he was friendly. Despite the difference in their ages, the marriage was a happy one; Lagrange found himself shopping and attending balls with his new bride. It was during this period that he recognized the need to perfect the calculus, and though his own attempt, *Theory of Analytic Functions Containing the Principles of the Differential Calculus,* was not too successful, it served as an inspiration for the nineteenth century mathematicians who made such advances in this field.

In 1799, Napoleon came to power, and the new French ruler lost no time in summoning Lagrange to his court. "Lagrange is the lofty pyramid of the mathematical sciences," he declared, as he made the mathematician a Grand Officer of the Legion of Honor. Lagrange accepted this new honor calmly, and returned home to continue his research.

After a brief illness, Lagrange died in Paris in 1813. He was buried in state in the Pantheon, near the tomb of Descartes. The French minister Tallyrand, speaking to Lagrange's father, had earlier paid the great man a fitting tribute: "Your son, whom France is proud to possess, has done honor to all mankind by his genius!"

VII

GAUSS

"ATTENTION, pupils! Today's problem is before you. You will work on it until the school day is over. Begin at once!"

The schoolmaster's harsh voice echoed through the cheerless room. Outside the smoke-blurred windows, an oak bough snapped in a frozen retort. Within, a pitiful fire in the heavy black stove gasped for breath. Herr Büttner's eyes moved slowly around the classroom, then halted suddenly.

"Gauss! You disobedient wretch! I'll teach you to start working when you are told!" The schoolmaster's coarse face hardened as he reached for the bullwhip on his desk. "Come here immediately!"

The startled look on young Carl Gauss's (GOUS) face turned to one of fright at the master's words. "But, Herr Büttner . . . I *did* begin when I was told. I have just finished, and you said . . ."

"Silence!" The angry master strode down the narrow aisle and seized the trembling student's ear. Then his expression changed; a sneer crossed his face. "So . . . you have finished. *Finished*!" He turned to the other students who had been listening without a sound. "Well! Our young genius has completed a problem which would take any grown man the greater part of an hour." The master's face reddened; his voice cracked with wrath. "Let us hear your brilliant solution then, Gauss, before I beat you twice as hard for your lies!"

The terrified boy backed away, clutching his chair for support. "You told us to add all the numbers from 1 to 100," he began in a trembling voice. "Well, instead of adding them in order, I decided it would be easier to start from both ends. $1 + 100 = 101$, $2 + 99 = 101$, $3 + 98 = 101$, $4 + 97 = 101$ . . . in fact, there are fifty pairs of 101's in the problem, ending with $50 + 51$. Since that is true, and since $50 \times 101 = 5050$, the answer to the problem is 5050."

Herr Büttner stared incredulously at Gauss, then seized the lad's writing tablet. Instead of the scrawled sums of the other students, only one number was visible there: 5050. The master's dull eyes bulged; slowly, dazedly, he shook his head. "All . . . all of you . . . except Gauss here . . . you are dismissed," he stammered at last. Then, uncertainly, he turned towards his desk.

The astonished students sat motionless for a moment, then bounded from the room. Only Carl remained rigid beside his desk. "Sir, please . . ." he murmured fearfully, watching Herr Büttner's contorted face.

The schoolmaster swung around. "You do not belong in this class, Gauss," he said in a strangely subdued voice. "From now on you will have private arithmetic lessons with my assistant, Bartels. The finest books will be provided for you. Since you and Bartels share the language of mathematics, you will make progress. Now you may go."

Poor Büttner! His shocked reaction was understandable. He could hardly have known that the boy before him, Carl Friederich Gauss, would one day be the giant of nineteenth century mathematics, a man whose genius is comparable to that of Archimedes or Newton!

An unpainted shack in Brunswick, Germany, was Gauss's first home. He was born there on April 30, 1777. Carl's parents were poor, uneducated people; his father worked at gardening and bricklaying. Although Herr Gauss thought education a waste of time, his wife encouraged her son's interest in learning.

The boy's talent for mathematics was evident from the first. He once remarked that he had learned to add and subtract before he could talk. At the age of three, while playing with a wooden toy one afternoon, he heard his father totalling the workers' payroll. "That's wrong, Papa," the child interrupted, "your sum should be sixty-three marks." Carl received a stinging blow for his correct answer, for Herr Gauss's ignorance was matched by his brutality.

Carl's Uncle Friederich, his mother's brother, was a different sort, however. A silk weaver by trade, he was an intelligent, kindhearted man. He observed his tiny nephew teaching himself to read, and began to bring the child books. Friederich's interest in Carl's education continued for years; his enthusiasm must have been a special source of encouragement to the boy.

One of young Carl's favorite pastimes was playing beside the tree-shaded canal near his home. As he pranced along the stone dike, he would occasionally peer into the murky water at his contorted reflection. One afternoon he leaned too far; the canal tender heard a feeble call and reached the boy just as he was about to drown. It was a close call—for Carl and for mathematics!

Gauss first attended school at the age of seven. In spite of Herr Büttner's unimaginative teaching, Carl found his new life exciting. He read constantly, often under difficult circumstances. When his father forbade him to "waste" the family's wood supply by reading near the fire, Carl would obediently go to his room—with a turnip under his arm. Then, cutting a hollow in the vegetable, he would insert a cotton wick, light it, and read for hours in the dimly lit, freezing bedroom.

Sometimes Carl would sneak into the Duke of Brunswick's palace gardens with a book under his arm. There he could find a quiet place to read without being disturbed by his father's scolding. One morning, as he was settling himself beneath a bush, a twig snapped on a nearby path. "I've caught you, you urchin," laughed the young woman who swept into view. Carl sprang up in a panic; before him stood the Duchess of Brunswick!

"Don't be afraid," the Duchess soothed. "I know that you often come here. What are you reading?"

Carl obediently told her about his book, and the Duchess continued to question him. Finally the noble lady smiled. "My husband, the Duke, will be interested to hear about his young visitor," she remarked. "You can expect to hear from him soon."

The next day, a glittering carriage rolled up to the door of the Gauss cottage. Georg Gauss, Carl's brother, took one horrified look at the servants who alighted and bolted for the kitchen. "They've come to take you to the palace," he screamed at Carl. "Now you'll get it, you good-for-nothing! See what happens when you always have your nose in a book!"

Georg couldn't have been more mistaken. The Duke had been so impressed by his wife's account of her meeting with Carl that he had decided to become Carl's benefactor. This was an unbelievable stroke of luck—it meant that the nobleman wished to pay for all of Carl's schooling, and to assist him in later life if necessary.

The Duke's confidence seems to have been the trigger which released Carl's genius. The boy was only fourteen when he first met the Duke. The next year, at the age of fifteen, he made a thorough study of Greek and Latin language and literature. At sixteen he thought of an alternative to Euclid's geometry, and put forth the astonishing and accurate idea that another, non-Euclidean geometry exists. In his seventeenth year, young Gauss correctly criticized proofs in the theory of numbers which had been taken for granted for centuries.

By the time Gauss was eighteen, he had mastered the works of Lagrange and Newton. His professors at the local college agreed that the young mathematician knew far more than they, and the learned men watched with relief as Gauss departed for Germany's famous University of Göttingen.

At Göttingen, Gauss made the discovery which is supposed to have set his feet firmly on the path of mathematics. He found a method of constructing a regular polygon of seventeen sides, using only ruler and compass. Gauss was understandably pleased with his accomplishment; he even asked that a model of the polygon be engraved on his tombstone when he died. (Although this was never done, there *is* such an engraving on a monument to Gauss in Brunswick.)

The crowning achievement of Gauss's university career was his doctoral thesis, written at the age of twenty. In this, he gave the first satisfactory proof of the fundamental theorem of algebra, which deals with algebraic equations. Newton himself had never been able to supply this proof.

In 1801, Gauss published his first major work, the *Disquisitiones Arithmeticae* (*Arithmetic Researches*). This early masterpiece on number theory brought him instant fame. Upon reading it, Lagrange wrote to Gauss, "Your work has raised you at once to the rank of the first mathematicians, and I regard the last section as containing the most beautiful analytical discovery that has been made for a long time."

Gauss's dedication of the *Arithmetic Researches* shows that his years of brilliant success had not gone to his head. He addressed the Duke: "If your Grace had not opened up for me the access to the sciences, if your unremitting benefaction had not encouraged my studies up to this day, I would never have been able to dedicate myself completely to the mathematical sciences . . ."

The influence of the *Arithmetic Researches* was enormous, for the material it contains is basic to modern number theory. Gauss considered this branch of mathematics to be of prime importance. "Mathematics is the queen of the sciences," he once commented, "and the theory of numbers is the queen of mathematics."

Oddly enough, Gauss turned his attention to astronomy when he completed the *Arithmetic Researches*. He had become interested in Ceres and Pallas, two small heavenly bodies, and for the next few years he did research in astronomy and electromagnetics as part of his investigation of their movements. He also let himself be distracted from scientific matters long enough to get married, at the age of twenty-eight.

In 1807, Gauss was named director of the University of Göttingen. During his first years as a member of the faculty, he occasionally lectured to the mathematics students. Teaching did not appeal to Gauss, however; he found the preparation irksome and the students poorly prepared.

Gauss was an unassuming, humble man. He could never understand why anyone would pretend to know something, or refuse to admit a mistake. "It is not knowledge, but the act of learning . . . which grants the greatest enjoyment," he wrote once. "I imagine the world conqueror must feel thus, who, after one kingdom is scarcely conquered, stretches out his arms again for others."

The mathematician conquered his kingdoms in a cramped study at Göttingen, furnished only with a work table, desk, chair, and lamp. Although he set forth his theories in a clear, orderly fashion, many of them could not be understood by his colleagues. Gauss was puzzled by this fact. "If others would reflect on mathematical truths as deeply and as continuously as I have, they would make my discoveries," he remarked to a friend.

Gauss's interest in Ceres and Pallas bore fruit in

1809 in the form of his second book, *The Motion of the Planets.* Riches and fame were showered upon him as a result, although other mathematicians were saddened that Gauss seemed to have temporarily neglected their branch of science. He had not done so, however; his diary, found after his death, recorded many brilliant mathematical ideas on which he had been working all along, which have proved essential to the development of modern mathematics.

In 1821, Gauss was appointed scientific adviser to a government geodetic survey. This work inspired new mathematical theories, particularly in relation to surfaces and mapping. As the years passed, Gauss turned his attention to mathematical physics and complicated research in geometry. Even in his old age he advanced startlingly original theories.

Little of Gauss's work was published during his lifetime, chiefly because of the mathematician's insistance on perfection in his publications. When a friend chided him for keeping the "infinite richness of your ideas" from the scientific community, Gauss replied, "I do not like to erect a building in which main parts are missing." "A cathedral is not a cathedral," he later added, "until the last scaffolding is down and out of sight."

Gauss's quiet manner hid a constantly active mind.

If in the middle of a conversation he happened to think of a new theory, he would instantly become lost in thought. His unbelievable powers of concentration allowed him to do most of his calculations mentally.

Although Gauss's magnificent work in mathematics was his main achievement, his interests were not restricted to science. His hobbies included the study of European literature, ancient classics, world politics, botany, and minerology. He could read and write Latin, English, Danish, and French, as well as his native German. At the age of sixty-two, he decided to add Russian to the list, and in two years was able to read and write this difficult language without a mistake, according to a Russian visitor to Göttingen.

Gauss showed a streak of inventive genius throughout his lifetime. The heliotrope, a device which transmitted signals by reflected light, was his invention. He also designed an electric telegraph in 1833, and sent messages on it regularly. In fact, his research in electromagnetism and related fields greatly aided the advance of physical science in the nineteenth century.

Gauss remained at Göttingen until his death in 1855, at the age of seventy-eight. His contribution to mathematics had been overwhelming. The astronomer Brendel wrote of this mathematical giant:

His mind penetrated into the deepest secrets of number, space, and nature;

He measured the course of the stars, the form and forces of the earth;

He carried within himself the evolution of mathematical sciences of a coming century.

VIII

GALOIS

"CAMILLE! Your book is a grand success! All the Academy members are praising it to the skies. And just read what the reviewer in the *Mathematical Journal* has to say!"

The exuberant young lady burst into the study, let her armful of newspapers and magazines drop, and whirled her husband around in a merry dance. The author joined in her delighted laughter, then stopped and looked suspiciously at her.

"But, Annette, what did they say about . . .?"

His wife tossed her head impatiently. "Camille Jordan, you are impossible! Very well, then, I won't keep you in suspense. All of them—the *Journal* reviewer, the Academy members—agree that your M. Galois is a genius, a superb mathematician, a marvel! Now are you content?"

Jordan's face broke into a relieved smile. "My dear, that is the best news you could have brought me. Galois' mathematical ideas are too important to be neglected any longer. It is for that reason that I included his work in my book. Now, perhaps, the mathematical world will give the fellow his due!"

Camille Jordan's prophecy was soon to come true. Mathematical thinking had finally become advanced enough to recognize Galois' genius, though it was forty years since the mathematician had died. His work had come to light at last—a handful of pages that was enough to place Galois among the torchbearers of mathematics!

Évariste Galois (gal-WAH) was born on October 25, 1811, in Bourg-la-Reine, a town near Paris. His father was a man of scholarly interests; his mother had been carefully educated by her father, a judge. Neither side of the family seems to have had any interest in mathematics, nor did Évariste show an early mathematical talent. His favorite childhood pastime was writing stories and poems.

In their desire to give Évariste the best education possible, M. and Mme. Galois made the unfortunate mistake of enrolling their son at Louis-le-Grand, a boarding school near Paris. The prison-like atmosphere of the school, and the harshness of its dictatorial headmaster, made Galois long for the happy village life at Bourg-la-Reine. Although his schoolwork showed promise at first, Évariste soon tired of the repetitive memorizing that was required of him. The only subject which interested the boy, arithmetic, was considered unimportant by the school staff.

As the years passed, however, Galois became more and more fascinated by mathematics. By the time he was fourteen, he had begun to pore over the works of Lagrange and the Norwegian, Abel. Soon Galois became completely absorbed in the writings of the great mathematicians. He daydreamed about their theories in class, forgot his books and his assignments, and annoyed his teachers by his inattention. "I think Galois' parents had better let him take only mathematics," one teacher finally reported. "The mathematical madness dominates this boy."

Galois had already come to the same conclusion. He decided to take the examination for the École Polytechnique, France's great school of science, where he felt his mathematical ability would be encouraged. So certain was he of success that he felt it unnecessary to prepare for the test. On the day of the examination, the problems proved as simple as he had expected, and he awaited his acceptance with confidence.

The results of the examination were posted soon afterwards. Unbelievably, Galois had failed! His unusual method of solving problems—in his head, without detail or explanation—had been his downfall. As one mathematician later remarked, "A candidate of superior intelligence is lost because of an examiner of inferior intelligence."

Disappointed but still optimistic, Galois returned to Louis-le-Grand. His new math teacher, M. Richard, immediately recognized the young man's genius. "This pupil has a marked superiority above all his fellow students," Richard wrote of Galois. "He works only at the most advanced parts of mathematics." The teacher spent much of his class time explaining Galois' findings to the other students, and vainly begged the Polytechnical School to admit his prize pupil without examination.

M. Richard had reason to be proud of Galois. The seventeen-year-old had already begun to make revolutionary changes in the theory of equations—changes whose effect on mathematics is still being felt. His was the realm of pure mathematics, and few men of his century could understand the sparks of genius which flew from Galois' pen.

In 1829, Galois published his first paper, on fractions. A few months later, he prepared a summary of all his mathematical findings and sent it to the French Academy, hoping for the support of that scholarly group. Unfortunately, the mathematician Cauchy, who received Galois' manuscript, must have been either jealous of or unimpressed by it. He first neglected to bring the paper to his colleagues' attention, then lost it altogether.

In desperation, Galois applied again to the École Polytechnique, knowing that a second failure would close its doors to him forever. This time, the examiners were ready for Galois; they had heard of the brash young man who dared to present his work to the Academy. Their taunting remarks and questions made it clear to Galois that he could not hope for acceptance, and, in a fit of rage, he hurled an eraser at one of his tormenters and ran blindly from the examination hall before the test was finished.

Only one path lay open to Galois now that his mathematical career seemed finished. In February, 1830, he entered the Preparatory School, a teacher-training institute. Galois had decided to try to combine a teaching career with independent mathematical research.

By this time, Galois was nineteen years old. As though he suspected that his life was nearly over, he tackled the theory of algebraic equations with brilliant ferocity. Fully realizing the importance of his research, Galois decided to enter a manuscript in the contest for the Grand Prize in Mathematics, sponsored by the Academy of Sciences. As he himself stated, quite accurately, "I have carried out researches which will halt many scholars in theirs."

The events which followed this fateful decision are still unclear. Galois' paper was speedily delivered to the Secretary of the Academy, who was just about to read it when he suddenly died. A shiver must have run through Galois' body when he learned this news; he immediately rushed to the Secretary's home to regain his manuscript. There, to his horror, he found that the paper had disappeared!

It seems impossible that this chain of events was the result of sheer coincidence. Could Galois have been the victim of a plot, concocted by envious mathematicians to keep him from outdistancing them? No one knows; in any case, the paper's disappearance drove Galois into a bitter state from which he never fully recovered.

In July, 1830, Paris was suddenly swept by the tornado of revolution. King Charles X had dissolved the Chamber of Deputies and suspended the Constitution, and the people rebelled at once. The revolution provided an outlet for Galois' frustrated spirit, and he began an attempt to organize his fellow students against the king's tyranny. However, the director learned of the plan, locked the students in their rooms, and expelled Galois.

The young mathematician immediately became a member of the Society of the Friends of the People, a Republican revolutionary group. He also joined the National Artillery Guard, and spent three days a week at the Louvre in military drills and shooting practice. Although Galois' new guard uniform was a splendid sight—blue jacket and cap with red trimming, and red striped trousers—it was to be another source of trouble for the unlucky fellow.

Galois had not abandoned mathematics during these chaotic months. In January, 1831, he began a short-lived career as a tutor of mathematics. At his first lecture, held in a bookshop, he gave his own view of scientific discovery. "The march of science is not along a straight road," he began. "Science develops along a strangely shaped path, and in its progress accident does not play a minor role . . . When creating, the scientist does not deduce; he combines, he compares. He does not arrive at the truth; he hits upon it as though by accident."

The opening lecture drew a crowd of forty listeners, including students, Republicans, and police spies. Not one mathematician appeared. By the third lecture, the audience had dwindled to four, and the course was cancelled. Once again, Galois' genius had been underestimated.

A few weeks later, for the third and last time, Galois submitted a new manuscript to the Academy of Sciences. Under its title, *On the Conditions for Solvability of Equations by Radicals,* appeared the mathematical masterpiece which is now called the Galois theory. Galois' introduction took the form of a plea: "This paper is a summary of a work which I had the honor to present to the Academy a year ago. Since this work was not understood, and since the propositions which it contained were undoubtedly questioned, I shall content myself with giving here . . . the general principles, and only one application of my theory. I appeal to the referees to read at least these few pages with attention."

While Galois was waiting for a reply from the Academy, he continued his revolutionary activities. On May 9, he attended a Republican banquet and protest meeting. Stirred by the ardor of the young radicals who were present, Galois proposed a toast—to the new king, Louis Philippe! Since Galois had a knife in his hand as he raised the glass, his action was interpreted by royalist spies as a threat against the monarch's life. Next morning, the police broke into Galois' room and arrested him.

After pacing up and down in a prison cell for a month, Galois was brought to trial. Luckily, the jury took pity on the young man and acquitted him. But no sooner had Galois returned to his home than he was arrested again, this time as a "dangerous radical." The new charge brought against Galois was that he had illegally worn the uniform of the Artillery Guard, which Louis Philippe had banned. At his second trial, Galois was sentenced to the prison of Saint-Pélagie for six months.

"He is already so tired," lamented Galois' sister after she visited him at Saint-Pélagie. "He abandons himself entirely to distressing thoughts. He is gloomy and old before his time. His eyes are hollow as though he were fifty years old."

The prison atmosphere was not the only factor which had aged Galois. A few days after being thrown into his cell, he had received the awaited letter from the Secretary of the Academy. M. Poisson had finally given his verdict of Galois' latest manuscript.

"We have made every effort to understand M. Galois' proofs," wrote the haughty Poisson. "They are incomprehensible. His argument is neither sufficiently clear nor sufficiently developed to allow us to judge its rigor; it is not even possible for us to give an idea of this paper. One should wait to form a definite opinion . . . until the author publishes a more complete account of his work."

Galois realized at once that ignorance guided M. Poisson's decision. In the introduction to *Two Papers on Pure Analysis,* written in prison, he commented, "In these two papers, the reader will find the remark, 'I don't know.' I am aware that by writing this I am exposing myself to the laughter of fools. Unfortunately, hardly anyone realizes that the most precious and scholarly books are those in which the author states clearly what he does not know, because an author harms his reader most by concealing a difficulty."

In March, 1832, near the end of his prison term, Galois was sent to a nursing home because of his run-down condition. There he met and took a liking to Eve Sorel, a Republican girl. The two spent many hours discussing French politics; outside, the dread disease cholera raged through the streets of Paris. After the epidemic had run its course, Galois was released.

A few nights later, Galois was accosted on the street by another Republican who claimed to be involved with Mlle. Sorel. Harsh words were exchanged, and the stranger challenged Galois to a duel the next morning. (Some historians suspect that the challenger was really a royalist, sent by the police to get rid of Galois.)

The mathematician must have sensed that the morning would bring fate's last blow. He rushed home and began at once to write down all the glorious mathematical ideas which he had never been able to develop fully, and which have challenged generations of mathematicians ever since. His brilliant treatment of the theory of groups was included in the hastily-written pages; it is but one example of the genius of Galois. The pathetic comment scribbled in the margin of one of the pages, "I have not time," makes one wonder how much more Galois would have contributed to mathematics if he had had the years of Newton or Gauss.

Galois added two letters to his hastily-written manuscript. In one, he begged his friend, Auguste Chevalier, "Preserve my memory, since fate has not given me life enough for my country to know my name." The other letter, to his Republican colleagues, lamented, "Oh, why do I have to die for such an unimportant cause? Farewell! It was my wish to give my life for the public good!"

The next morning was gray and chill. The combatants began their somber pacing at 5:00 A.M. A few moments later, Galois fell writhing to the ground. The next day, May 31, 1832, the young mathematician died.

Galois' body was buried in a common grave; today

no one knows where it is located. His work fared better, for Chevalier was a devoted friend. He saw to it that Galois' last manuscript was printed in the *Revue Encyclopedique* later that year. Although it was overlooked at the time, a later mathematician, Liouville, sensed its worth and had the paper republished in the *Mathematical Journal*. Finally, in 1870, another mathematician named Jordan included Galois' theories in his own book. Jordan's work became famous; Galois' ideas became immortal.

The final scene in the story of Galois took place seventy-seven years after his death. On June 13, 1909, there was a splendid ceremony at Bourg-la-Reine, in honor of the town's neglected son. Present were the mayor, the Secretary of the French Academy, famous mathematicians, and important French officials. The Director of the Preparatory School read a touching speech which concluded, "We are here to make honorable amends to the genius of Galois in the name of the school that he entered with regret, where he was misunderstood, from which he was expelled, and of which he is one of the most stirring glories."

IX

VON NEUMANN

"I DON'T THINK I'd like to be a mathematician in the twentieth century," a boy or girl will sometimes remark. "Other scientists are doing exciting things, like launching rockets or conducting undersea explorations, but mathematicians just sit around working on a lot of numbers."

How amused John von Neumann (von NOY-mahn) would have been by that comment! His career as one of the most brilliant mathematicians of this century included work on the atomic bomb and ballistic missiles, a study of poker and other games of chance, and leadership in the development of high-speed electronic computers, as well as magnificent contributions to the field of pure mathematics. It could hardly have been called dull!

Von Neumann was born in Budapest, Hungary, on December 28, 1903. As a child he showed extraordinary ability in mathematics; at the age of six he could divide two eight-digit numbers in his head, and he began the study of college calculus a few years later. Those who knew Jancsi von Neumann called him *Wunderkind*—"Wonderchild."

Jancsi's classmates were more impressed by their companion's photographic memory than by his mathematical achievements, however. One afternoon they watched in amazement as he flipped through the telephone book, glanced at several of the pages, and promptly recited from memory the names, addresses, and phone numbers on each page! The boy's teachers had had a similar surprise when they saw their ten-year-old pupil read forty-six volumes of a book of German history, and then discuss the work in detail with Hungarian military leaders.

Max von Neumann had always hoped that his son would join him in business as a banker. He soon realized that Jancsi had a different future ahead of him, however, and he enrolled the young student at the Lutheran High School in Budapest. Jancsi distinguished himself in all his subjects; his superior work attracted the attention of a renowned Hungarian mathematician named Leopold Fejer. Fejer gave Jancsi special instruction in mathematics, and was later to call him, "Our Country's Greatest Jancsi!"

The years that followed high school took John von Neumann from Budapest to Germany and Switzerland in search of a thorough scientific education. From 1921-1923, he studied chemistry at the University of Berlin. Two years later, he received a diploma in chemical engineering from the Technical High School in Zurich, Switzerland. Finally, in 1926, he was awarded the doctorate in mathematics from the University of Budapest.

No sooner had the young mathematician finished his studies at Budapest than he was named to a Rockefeller fellowship at the University of Göttingen. There, at the age of twenty-three, von Neumann published his first masterpiece, *Mathematical Foundations of Quantum Mechanics.* The work was vital to the understanding of the quantum theory, on which all atomic and nuclear physics is based. (The quantum theory holds that the taking in or giving off of energy by atoms or molecules is not continuous, but occurs in distinct amounts, called quantums.)

In 1927, von Neumann accepted his first teaching position, at the University of Berlin. The new instructor's teaching methods raised many eyebrows; he lectured without notes, and chose for his classes problems which he had not yet solved, so that he might attempt their solution along with the students.

It was in Berlin that John von Neumann began a study of the game of poker. This game in particular intrigued him, because it involved not only the factor of chance (which is present in a dice game, for example), but also the question of the player's strategy. Could such a game be defined in mathematical terms?

The young mathematician set to work. Within a few months he had developed a brilliant "theory of games," which brought into being a new field of mathematical study. Von Neumann's approach could be applied not only to games of chance and strategy, but also to such important areas as economics, military strategy, and sociology.

Although the "theory of games" was first published in 1928, when von Neumann was only twenty-five years old, it was immediately recognized as a mathematical work of art. No important changes have since been made in its basic ideas. The theory received further acclaim when it was applied to economic structures by the economist Oskar Morgenstern in 1944. He and von Neumann published the results that year in the book, *Theory of Games and Economic Behavior.*

By 1930, von Neumann sensed that Europe was moving towards tragedy. He had accepted an offer to lecture for a year at Princeton University in New Jersey. The mathematician was stimulated by Princeton's intellectual atmosphere and charmed by its quiet beauty. When in 1931 he was offered a professorship of mathematics and physics at the university, he decided to remain in the United States.

Von Neumann's unorthodox teaching attracted attention in Princeton as well as in Berlin. Many of his students complained that he was simply too brilliant to teach, that he passed too quickly from one mathematical step to another, leaving the class in confusion. Even some of his fellow professors were startled by the ease with which he solved extremely complex problems. One of them, watching von Neumann clear the blackboard again and again to make room for the successive steps of a difficult problem, remarked with a chuckle, "I see. Proof by erasure!"

In 1933, von Neumann was invited to become a full professor at the Institute for Advanced Study, a newly opened international center for scholars in Princeton. He immersed himself there in mathematical research for several years. Then World War II struck, and the new American citizen was suddenly called upon to act as military consultant to the United States government.

Soon von Neumann was tackling problems of submarine warfare and atomic weapons. His work on the problem of detonation of the atomic bomb brought the Los Alamos project to completion a year earlier than was thought possible. For his outstanding contributions to the American military effort, von Neumann was awarded the Medal for Merit and the Distinguished Civilian Service Award in 1946.

After the war, von Neumann resumed the life of a mathematician—*his* kind of mathematician. Far from being devoted to his work alone, he relished lively parties and new cars. One of his favorite hobbies was writing poetry; he delighted in thinking up nonsensical limericks, and could make a pun in three languages.

The mathematician's greatest joy lay in the challenge of his work, however. Von Neumann had become fascinated by the possibilities of calculating machines before the war; he was now about to launch a full-scale exploration of their potential which would eventually make him the foremost expert on computers in the United States.

As the first steps in his computer research, von Neumann studied psychiatry, consulted neurologists, and discussed the human brain with hundreds of experts. Then, under the mathematician's watchful eye, the design of his computer took shape. At last the job was done; von Neumann's complex creation was unveiled.

MANIAC (the initials stand for mathematical analyzer, numerical integrator, and computer) served as a model for many future computers. Its calculations were considered awesomely swift at the time; the machine could do two thousand multiplications a second, and complete in one hour a problem which had formerly taken several years!

NORC (naval ordinance research computer) was von Neumann's second computer. This talented machine could give a twenty-four-hour weather forecast in a few minutes' time, record information about the earth's core, figure the tidal motions of the Atlantic and Pacific Oceans, and solve military maneuver problems. The Rand Corporation had a copy of NORC; it was appropriately called JOHNNIAC.

When asked his reason for devoting so much time to computer development, von Neumann gave the reply of a true scientist. "Problems which seem interesting to scientists, but which formerly were impossible for practical reasons, can now be undertaken by this new tool, thus widening the range of scientific investigation," he explained.

Before long, scientists were using ORDVAC (ordnance variable automatic computer) to solve complicated problems of analysis; UNIVAC (universal automatic computer) was flashing election returns to the American public. The electronic brains which von Neumann had such an important role in developing brought undreamed-of changes to twentieth-century life; most of the dazzling scientific, educational, and industrial advances of our time would not have been possible without them.

Skeptics sometimes wondered if the new computers would eventually take the place of the human brain. Von Neumann had his own opinion on that subject. "I know what extraordinarily complicated machinery the human intelligence can devise," he wrote. "I also know something of the complications of the human nervous system. There is no comparison between the human nervous system and the most complicated machine that human intelligence has ever devised, or can ever devise."

Anyone who has ever doubted the capacity of the human brain should be reassured by the following story. One afternoon, von Neumann received a desperate telephone call from a research corporation, telling him of a problem so difficult that no known computer could solve it. The corporation executive begged von Neumann to design a machine advanced enough to tackle this intricate question.

Von Neumann agreed to meet with the corporation scientists, requesting only that an explanation of the problem itself be presented. So complex was the situation that the scientists took several hours merely to give an adequate description. After they finished, there was a short pause.

"Well, gentlemen, you won't need a computer," von Neumann remarked at last, jotting down several numbers. "Here is your answer. Now, where can we get some lunch?"

Perhaps, as the physicist Edward Teller said, von Neumann's uncanny ability was "due to an early discovery of the joy of thinking and an almost perpetual exercise of this activity." The amusing "railroad story" seems to prove Teller's point. One morning a friend was taking von Neumann to catch a train for Chicago. On the way, the man pulled a sheaf of papers from his pocket and laughingly commented, "Here's something to keep you busy for the next few hours." The "something" was a devilish math problem which had been solved only once, after a week of constant study, by a brilliant Russian mathematician who was journeying across Russia by train.

If von Neumann's companion noticed a gleam in the mathematician's eye, he thought little of it at the time. Two days later, however, he was stunned to find in his mailbox a thick envelope from Chicago containing a fifty-page handwritten solution to the problem! Von Neumann had added a P.S.: "Running time to Chicago—15 hrs., 26 min."

Because von Neumann often became totally absorbed in his mathematical ideas, he was accused of being absentminded. Colleagues exchanged amused glances as he searched his pockets for money or plane tickets which were usually to be found on his bureau back in Princeton. Several times he became so preoccupied with a problem while traveling that he had to call his wife to find out why he had taken the trip in the first place! On the other hand, he could summon amazing concentration when necessary; he did much of his work while listening to music in his living room, and regularly worked out problems in crowded restaurants or at noisy parties.

In 1953, von Neumann was named chairman of a committee of scientists and military leaders who were attempting to evaluate America's missile program. Under his leadership, work on the intercontinental ballistic missile (ICBM) was begun. Participants on the committee spoke with awe of von Neumann's talent in handling people, as well as his ability to get to the core of a discussion. His agile mind leapt from one idea to another; one of the officers present recalls, "None of the specialists nor experts ever discussed any phase . . . of their individual fields but that von Neumann seemed to know all about them."

During the development of the ICBM project, a physicist who had been experimenting on one phase for nearly a year asked von Neumann's opinion of his work thus far. The mathematician thumbed slowly through the stack of papers, then thought for a few moments. "It won't work," he finally answered. Looking dubious, the physicist returned to his laboratory to review the project. After two months of careful study, he realized that von Neumann had been right.

The busy months of government work left von Neumann little time for purely mathematical pursuits. He sometimes snatched a few minutes of relaxation with his frolicsome dog, Inverse, or took an evening off to read about his hobby, medieval history. He could never seem to keep away from mathematics, however, and even amused himself on automobile trips by looking for prime numbers on license plates.

In 1954, von Neumann was appointed to the top-level Atomic Energy Commission. Though the new position would force the mathematician to move to Washington, D. C., and severely limit his time for research, von Neumann accepted the post immediately. "All scientists whose work has made them familiar with atomic energy matters have an obligation to take a turn in shouldering the administrative responsibility," he explained to newspaper interviewers.

A few months after von Neumann moved to Washington, he became seriously ill, and was told that he had only a short time to live. The mathematician threw himself into his work, vainly trying to finish one or two projects. It was no use; in February, 1957, John von Neumann died.

As one reviews von Neumann's phenomenal accomplishments, it seems impossible to believe that they are the product of one man's mind. Perhaps the words of the physicist Hans Bethe offer the best explanation of von Neumann's genius. He wrote, "I have sometimes wondered whether a brain like von Neumann's does not indicate a species superior to that of man."

X

WIENER

A CHEERLESS New England sun gazed bleakly down on the icy road adjoining the college campus. The solitary figure of an eleven-year-old boy trudged slowly up the incline, pulling his sled in uneven jerks. Ruts worn in the snow by the runners testified to a busy morning of sledding.

A noise of crunching footsteps behind him made the youngster pause and glance uneasily over his shoulder. His wind-chapped face paled, and he broke into a clumsy, desperate run. Then came the hated chorus of popping flashbulbs and taunting calls. "Hey, Norbert!" "Say, genius, how about an interview?" "C'mon, college man, say something brilliant!"

As usual, Norbert Wiener's (WEE-ner or VEE-ner) panic helped him to outdistance the shouting reporters and photographers. In a matter of minutes he reached the mathematics building, breathless and angry. His day of sledding was ruined! Still, he thought, his face brightening, perhaps he could get in some research on the Galois theory before his afternoon physics class.

Such scenes were not unusual in the life of Norbert Wiener. His was the painful and confusing boyhood of a child genius—a prodigy. Unlike many such prodigies, however, Wiener became a brilliantly successful adult who enjoyed a long career as one of America's most eminent mathematicians.

Wiener was born in Columbia, Missouri, on November 26, 1894. While he was still an infant, his family moved to Massachusetts, where his father hoped to find a college teaching position. Eventually the elder Wiener became a Professor of Slavic Languages at Harvard University.

The influence of Leo Wiener on his son was extraordinary. Early in the boy's childhood, he had decided to take charge of Norbert's education. When he saw how quickly the boy responded, he set up a study schedule so demanding that most children would have collapsed under it. Norbert was physically as well as mentally strong, however, and the child's interest in learning, coupled with fear of his father, drove him to amazing intellectual achievements.

It is probable that young Norbert would have been an outstanding student even without his father's help. He had lisped the letters of the alphabet when he was a year and a half, and he was able to read by the age of three. At five, he enjoyed *The Arabian Nights* and *Alice in Wonderland.* Nevertheless, playing with a brightly-painted model of a battleship was his favorite pastime.

Norbert's school performance must have caused his teachers to throw up their hands in despair. In some ways, he seemed far below average—his handwriting was barely legible, he counted on his fingers, and he had trouble learning the multiplication tables. At the same time, he read books on physics, chemistry, and natural history with ease, performed complicated science experiments, and spoke German.

By the time Norbert was seven, his father had become impatient with the school program. He withdrew his son from school altogether, and for the next three years tutored Norbert daily in algebra, geometry, Latin, and German. Professor Weiner was an exacting and impatient teacher; one childish mistake would send him into fits of rage, and the lessons often ended in tearful scenes. Yet Wiener felt little bitterness toward his father in later years, mainly because he realized that the older man was only trying to set high standards for him.

In many ways, Norbert was an average boy. He and his neighborhood friends built snow forts, climbed fences, and entered contests at the local playground. Young Wiener had scientific hobbies, too: he pored over science kits and delighted in taking electric motors apart. Books which he read for fun included *The King of the Golden River,* and Jules Verne's *The Mysterious Island* and *Twenty Thousand Leagues Under the Sea.*

Norbert's lack of muscular control was evident in his ballplaying as well as in his handwriting. At the age of eight, however, he became even more awkward than usual, and started stumbling over his reading. His alarmed parents consulted a doctor, who found that Norbert had become seriously nearsighted, and forbade him to read for six months.

Professor Wiener decided to continue his son's lessons—by ear! Although this must have been a terrible strain on Norbert, he later recalled, "This period of ear training . . . was probably one of the most valuable disciplines through which I have ever gone, for it forced me to be able to do mathematics in my head, and to think of languages as they are spoken rather than as mere exercises in writing." Wiener's already unusual memory was sharpened during this difficult time, too.

In the summer of 1903, when Norbert was nine, the Wiener family moved to a farm in the Massachusetts countryside. There the boy and his fox terrier roamed through the marshy meadows and caught frogs and turtles in a nearby stream. The fun of sliding in the old barn's hayloft or rowing on the willow-fringed pond partly made up for the loss of Norbert's city friends.

That fall, because of his father's busy schedule, the nine-year-old was enrolled as a special student at Ayer High School, first as a freshman, then as a junior. Norbert had never attended school regularly, and he felt strange and out-of-place there. Luckily, students and teachers alike treated him with kindness and understanding. He was graduated from high school in 1906, at the age of eleven.

Although Norbert was now mentally ready for college, he was still a boy at heart. With a good friend, he built toy trains, exploded firecrackers, and made an electrical bell. His experiments with two neighborhood friends were even more ambitious. "We almost blew ourselves up by trying to make an internal combustion engine out of a tin fly spray, and we nearly shocked ourselves to death in amateur radio experiments," reminisced Wiener in his autobiography.

In September, 1906, Norbert was admitted to Tufts College in Massachusetts, and assigned to classes in Greek, German, English, special mathematics, physics, and chemistry. This marked the beginning of a painful period for the eleven-year-old. Hounded by news-hungry reporters who considered him a freak, and lonely for the companionship of boys his own age, Norbert spent most of his free time in the Children's Room of the Boston Public Library. He graduated from Tufts with honors in mathematics in three years, at the age of fourteen.

Norbert's father immediately enrolled him at the Harvard Graduate School, so that he might study for a doctorate in biology. There the boy's nearsightedness and clumsy movements made him botch the delicate laboratory work required of him. Professor Wiener promptly switched his son to Cornell University, where Norbert felt himself to be even more of a misfit. He returned to Harvard in 1911, and received his doctorate in Philosophy there two years later, at the age of eighteen.

In his autobiography, Wiener comments on these dark years: "I was miserable. I was ill at ease with my associates in the middle twenties, and there were no youngsters of my own age to replace them. [I had] . . . the problems of an undeveloped boy, spending his first long visit away from home."

Fortunately, a taste of the independence and companionship which he longed for awaited the unhappy teenager. He had won a fellowship to Cambridge University in England, and he spent the next year there in the company of learned mathematicians and sympathetic new friends. From Cambridge, he went on to Göttingen, where he studied under the great German mathematician, Hilbert. Wiener's next semester at Cambridge was cut short by the beginning of World War I, and he returned to the United States to finish the year at Columbia University in New York.

Wiener felt that the years abroad contributed immensely to his growth as a mathematician. "Mathematics is not only a subject to be done in the study but one to be discussed and lived with," he wrote, adding, "The effective mathematician is likely to be a powerful factor in changing the face of society. He does less than his full duty if he does not face this fact."

The young student remembered his stay in Göttingen as a time of social development as well. "I learned to meet people both like me and different, and to get along with them," he commented. At Göttingen he was also an enthusiastic participant in hiking and mountain-climbing expeditions similar to those which he had enjoyed during previous summers in the New England hills.

In 1915, Wiener began teaching philosophy at Harvard—to students about his own age. His major interest lay in mathematics, however, and the next year he became an instructor in that department at the University of Maine. By 1917, America's involvement in World War I had become a reality, and Wiener volunteered at once. He was unhappy to learn that his eyesight made any thought of military service impossible.

Since normal university life was temporarily halted, Wiener decided that his training could be put to best use in engineering work. He had no sooner begun such a job, however, when his father demanded that he leave to become a writer for *Encyclopedia Americana*. This in turn gave way to work on ballistic devices at the Aberdeen Proving Ground in Maryland.

The end of the war brought Wiener the most important appointment of his career. He was invited to join the Mathematics Department of the Massachusetts Institute of Technology in 1919. He accepted, became an instructor at MIT that year, and remained to become one of the Institute's most illustrious faculty members.

The young mathematician's early research at MIT took the form of brilliant and varied excursions in the realm of pure mathematics. The problems in motion and probability which he began to study at that time led him ultimately to his masterpiece—cybernetics. "The whole background of my ideas on cybernetics lies in the record of my earlier work," Wiener wrote.

Wiener defined cybernetics as "the theory of communication and control in the machine and in the living organism." The term itself came from the Greek word for steersman. One example of the type of mechanism which the mathematician concentrated on is the ordinary thermostat, which shuts off automatically when a certain heat level is reached. For his pioneer work in this complicated and important field, Wiener has been called "the father of automation."

Realizing that the application of his theory could lead to "the automatic factory" and possible unemployment, Wiener met with government and union leaders to discuss this problem. As one admirer said of him at the time, "He keeps seeing thousands of things that could be done in science, and he is always aware of the moral, social, and industrial consequences of his findings."

Wiener's work did not keep him confined to the MIT campus, for he recognized the need for a constant exchange of ideas with other mathematicians. He traveled widely throughout his lifetime, learning and lecturing in England, France, Germany, Switzerland, China, Japan, and India. In 1944, he left MIT for Mexico to begin a joint project with Dr. Arturo Rosenblueth on "the application of modern mathematical techniques to the study of the nervous system." Wiener hoped that his research in this area could be used to improve the iron lung, translate words into impulses for the deaf, and equip artificial arms and legs with a substitute for the lost sense of touch. His later study of brain waves was an even more significant contribution in this area.

During World War II, Wiener was instrumental in the design of fire-control apparatus for anti-aircraft guns. An amusing story is told in connection with this project. One day, at the Philadelphia Navy Yard, one of the huge guns began shooting erratically. The officier in charge wrote a frantic letter to Wiener at MIT, describing the gun's strange new pattern. As soon as he received the letter, Wiener phoned the unhappy officer. "If you look inside the gun's mechanism at exactly the point I will give you," he suggested, "you will find that a mouse has crawled in there and died." Needless to say, the mathematician was correct!

At MIT, Wiener was affectionately called "Doc" by the mathematics students, who were amused by his habit of throwing peanuts in the air and catching them in his mouth. Many of the men who studied under him recall his excited cry of mathematical discovery, "Hot stuff, boys!" And at least one graduate has a vivid memory of Wiener walking through the MIT corridors with his finger on the wall for guidance, so absorbed in his thoughts that when he came to an open classroom door, he unconsciously entered, circled the room, and left without ever realizing that he had been there!

Wiener himself tells of getting a flash of inspiration one night in a movie theater. "It simply distracted all my attention from the performance," he relates, "and I promptly left the theater to work out some of the details of my new plan." Even more astonishing is his account of a siege of pneumonia. Delirious with pain, he was at the same time aware of a math problem that had been bothering him for months. He reports that his struggle for health seemed to merge with his mathematical struggle, and that when he had recovered from the pneumonia, the problem was well on its way to solution.

In spite of international demands on his time, Wiener managed to live quietly in Massachusetts and New Hampshire with his wife Margaret and their two daughters. Hiking remained one of his favorite pastimes as he grew older. A further source of relaxation for the mathematician was the reading and writing of detective stories.

Wiener's publications were many and varied. His most important book, *Cybernetics,* created a stir among laymen as well as mathematicians when it appeared in 1949. In addition to dozens of brilliant mathematical works, he is known to the public as the author of *Ex-Prodigy, I Am a Mathematician, The Human Use of Human Beings,* and *God and Golem, Inc.*

In explaining his decision to write *I Am a Mathematician,* Wiener comments, "I mean to give to the reader . . . at least a hint of the thrill of mathematical creation . . . The rewards are of exactly the same character as those of the artist. To see a difficult material take living shape and meaning is to be Pygmalion, whether the material is stone or hard, stonelike logic."

In 1963, the aging mathematician won the National Medal of Science for his contribution to mathematics, engineering, and biological science. It was universally agreed that Wiener's special genius had made it possible for him to produce masterpieces in both mathematical theory and applied mathematics. Yet, as one critic commented, he was not only a scientist, but a philosopher, a linguist who spoke six languages, and a man thoroughly acquainted with art and literature.

Norbert Wiener died in Stockholm, Sweden, on March 18, 1964. His whole life had been a testimony to the courage and striving of the human mind and spirit. "It is the battle for learning which is significant, and not the victory," Wiener had written. Yet his own exciting achievements proved to be among the most victorious of the twentieth century.

INDEX

Abel, Niels Henrik, 81
Adelard of Bath, 10, 33
Alexander the Great, 5
Algebra: defined, 10; equations, 73, 84; origin of word, 34; theorem of, 73
Al-Khwarismi, 29-34
Al-Mamun, Caliph, 10, 32-34
Analytical Mechanics (Lagrange), 62, 63
Anti-aircraft guns, 112-113
Archimedean screw, 15
Archimedes, 12-23, 59
Arithmetic Researches (Gauss), 73-74
Aryabhatta, 24-29
Aryabhattiyc (Aryabhatta), 26-29
Atomic bombs, 91, 95-96
Atomic Energy Commission, 101
Automation, 111

Ballistic missiles, 91, 100
Barrow, Dr. Isaac, 49
Bartels, Herr, 69
Berlin Academy, 61
Bernoulli (mathematician), 55
Bethe, Hans, 102
Bhaskara, 28
Brahmans, 26
Brendel, Martin, 77-78
Brunswick, Duchess of, 71-72
Brunswick, Duke of, 71-72, 74
Buddha, 25-26
Büttner, Herr, 67-69, 71

Calculation of Integration and Equation (al-Khwarismi), 33-34
Calculus, the, 49, 50, 52-53
"Cattle Problem, The," 19
Cauchy, Augustin Louis, 83
Chanute (French ambassador), 42
Charles X, King, 84
Chevalier, Auguste, 89, 90
Christine, Queen, 42-43
Cicero, 20
Computers, electronic, 91, 96-99
Conon (Alexandrian astronomer), 19
Cybernetics, 111, 114
Cybernetics (Wiener), 114

Dandapani, Prince, 25
Data (Euclid), 10
Decimal number system, 24-34
Descartes, René, 35-44, 50
Discourse on the Method of Rightly Conducting the Reason and Seeking Truth in the Sciences (Descartes), 40-41
Disquisitiones Arithmeticae (Gauss), 73-74
Distinguished Civilian Service Award, 96

École Polytechnique, 65, 81-83
Electromagnetism, 77
Electronic computers, 91, 96-99
Encyclopedia Americana, 110
Equal sign, 38
Equations: algebraic, 73, 84; defined, 41; Lagrange's, 62; theory of, 41, 51
Euclid, 3-11, 59
Euler, Leonhard, 60
Ex-Prodigy (Wiener), 114

Fejer, Leopold, 93
Fibonacci, Leonardo, 33
Frederick the Great, 61

French Academy of Sciences, 60, 83, 84, 86, 87-88, 90
French Revolution, 63

Galileo, 50
Galois, Évariste, 79-90
Gauss, Carl Friederich, 67-78
Gauss, Georg, 72
Gautama, Prince, 25
Geometry: analytic, 41; basis of modern, 41; defined, 7; solid, 8-9
Geometry (Descartes), 41, 49
God and Golem, Inc. (Wiener), 114
Gram, the, 64
Groups, theory of, 89
Gupta emperors, 25

Halley, Edmund, 59
Heiberg, J. J., 17-18
Heliotrope, the, 77
Hieron, King, 12-15, 21
Hilbert, David, 109
Human Use of Human Beings, The (Wiener), 114

I am a Mathematician (Wiener), 114
ICBM (ballistic missile), 100
Institute for Advanced Study, 95

James II, King, 53
JOHNNIAC (computer), 97
Jordan, Annette, 79-80
Jordan, Camille, 79-80, 90

Kepler, Johannes, 50
Koti, the, 25-26

Lagrange, Joseph Louis, 57-66, 73-74, 81
Laws: of motion, 51; of universal gravitation, 49
Leibniz, Gottfried, 52, 55
Liouville, Joseph, 90
Liter, the, 64
Louis Philippe, King, 86-87
Louis XVI, 62

MANIAC (computer), 97
Marcellus, 22-23
Marie Antoinette, 62
Mathematical Foundations of Quantum Mechanics (von Neumann), 93
Mathematical Principles of Natural Philosophy (Newton), 50-51
Measurement of the Circle (Archimedes), 16-17
Medal for Merit, 96
Medal of Science, 114
Meter, the, 64-65
Method (Archimedes), 18-19
Metric system: acceptance of, 65; base of, 64-65; establishment of, 63-64
Minus sign, 38
Missiles, ballistic, 91, 100
Morgenstern, Oskar, 94
Motion, laws of, 51
Motion of the Planets, The (Gauss), 76
Muhammed ibn Musa, 32

Napoleon I, 66
Newton, Isaac, 45-46, 59, 73
NORC (computer), 97
Numbers: Brahmi, 27; cube root of, 28; decimal system, 24-34; Hindu-Arabic, 33; position of, 27-29; Roman, 33; square of, 28; syllable-, 28; theory of, 9

On the Conditions for Solvability of Equations by Radicals (Galois), 86
On the Divisions of Figures (Euclid), 10-11
On Floating Bodies (Archimedes), 19
On Levers (Archimedes), 19
On the Making of the Spheres (Archimedes), 20-21
On Mirrors (Archimedes), 19
On the Sphere and the Cylinder (Archimedes), 19
On Spirals (Archimedes), 19
ORDVAC (computer), 97

Phaedius, 13
Pi (π), 16-17
Plato, 5
Plus sign, 38
Plutarch, 15, 16

Poisson, M., 87-88
Poker, mathematical study of, 94
Pope, Alexander, 55
Positional number system, 27-29
Principia (Newton), 50-51
Proportions, theory of, 9
Ptolemy, King, 3-6, 14
Ptolemy (astronomer), 20
Pythagoras, 7-8
Pythagorean theorem, 7-8

Quantum theory, 93

Recorde, Robert, 38
Revue Encyclopedique (Galois), 90
Richard, M., 82
Roman numerals, 33
Rosenblueth, Dr. Artura, 112
Royal Swedish Academy of Sciences, 43

Sand Counter, The (Archimedes), 19
Society of the Friends of the People, 85
Solar system, 51
Sorel, Eve, 88
Spectrum, the, 49
Syllable-numbers, 28
Syracosia (ship), 14

Talleyrand-Perigord, Charles Maurice de, 66
Teller, Edward, 99
Theorems: of algebra, 73; of Archimedes, 18; Pythagorean, 7-8
Theories: of equations, 41, 51; Galois, 86; of groups, 89; of proportions, 9; quantum, 93
Theory of Analytic Functions Containing the Principles of the Differential Calculus (Lagrange), 66
Theory of Games and Economic Behavior (von Neumann and Morgenstern), 94
Thirteen Books of Element (Euclid), 6-11, 33, 48-49
Tsu Ch'ung-chih, 17
Two Papers on Pure Analysis (Galois), 88

UNIVAC (computer), 97
Universal Arithmetic, The (Newton), 51
Universal gravitation, law of, 49
University of Scholars, 6

Volume, Archimedes' discovery, 18-19
Von Neumann, John, 50, 91-102
Von Neumann, Max, 92

Wiener, Leo, 104-108, 110
Wiener, Margaret, 114
Wiener, Norbert, 103-115
William of Orange, 37
World, The (Descartes), 42

Zero, 33